BELLAIRE PUBLIC LIBRARY
BELLAIRE, OHIO

The Modeler's Manual

Other books by Robert Schleicher

Building and Flying Model Aircraft
Building and Displaying Model Aircraft
Model Railroading Handbook, vols. I and II
Tyco® Model Railroading Handbook
Dollhouses and Dioramas
Model Car, Truck and Motorcycle Handbook
Model Car Racing
The ETV Model Book

The Modeler's Manual

ROBERT SCHLEICHER

Trains

Planes

Ships

Military Vehicles

Cars

Rockets

CHILTON BOOK COMPANY   Radnor, Pennsylvania

Copyright © 1981 by Robert Schleicher
All Rights Reserved
Published in Radnor, Pennsylvania, by Chilton Book Company
and simultaneously in Scarborough, Ontario, Canada,
by Nelson Canada Limited

Library of Congress Catalog Card No. 80-70352
ISBN 0-8019-6996-4 hardcover
ISBN 0-8019-6997-2 paperback

Manufactured in the United States of America

FRONT PANEL OF COVER
Clockwise from top left: 1. Etched-metal precision kit wire wheels on a Matchbox M.G. TC. 2. Mike Czibovic's medieval castle; from a paper model imported by John Hathaway. 3. The Enzmann Starship; rear-projection photo by Kevin Atkins. 4. Revell's 3-foot-long plastic *Thermopylae;* rigging lines are included in kit. 5. Dan Wilson's 1/35-scale model of Japanese "Ho-Ha" Type 1 half-track; scratchbuilt plastic body, kit track and wheels. 6. Albert Hetzel's 1/48-scale Santa Fe Railway 2-10-2; built from sheets and rods of brass. 7. Lloyd Jones' scratchbuilt 1/32-scale Loening 0A-1A all-plastic biplane.

BACK PANEL
Clockwise from top left: 1. Lou Roberts' easy-to-build Peanut scale model. 2. Mike Czibovic's 1/700-scale waterline models include scratchbuilt American and Japenese warships made from sheet plastic. 3. Three hand-carved all-wood Grand Prix cars, based on the plans for the 1964 Ferrari, rolling in a Pinewood Derby race.

1 2 3 4 5 6 7 8 9 0   0 9 8 7 6 5 4 3 2 1

Contents

1. *Model Building*   1

                                              Scratchbuilding   1
                                              Kit Conversions   2
                                              Scale Models   3
                                              Parts Building   5
                                              Researching Model Prototypes   5
                                              Working with Plans   7
                                              Model-Building Tools   10
                                              Painting   11
                                              Applying Decals   14

PART I
Model-Building Projects

2. *Rockets and Spacecraft*   19

                                              The Perfect Fantasy   19
                                              Do-It-Yourself Designs   20
                                              Assembling Plastic Kits   23
                                              The Enzmann Starship Model   24
                                              Making Rockets that Fly   24
                                              Painting a Rocket   30
                                              Researching Rockets and Spacecraft   31

3. *Aircraft*   32

                                              Static vs. Flying Models   32
                                              Accurate Replicas   33
                                              Display Models   34
                                              Scratchbuilt Aircraft   34
                                              The Loening OA-1A   35
                                              Flying Aircraft   39
                                              Flying Stick Models   41

4. *Automobiles*   48

                                              Plastic Model Kits   48
                                              Metal Auto Miniatures   49
                                              Car Conversions   50
                                              The Automobile Museum   54
                                              Hand-Carved Car Bodies   54
                                              The Maserati 5000 GT Model   54
                                              Clear Plastic Bodies   56
                                              Vacuum-Formed Bodies   58

5. Ships   63

The Drydock   63
Ships that Sail   64
Assembling Model Ship Kits   67
Ship Kit Conversions   68
Scratchbuilding Waterline Models   68

6. Military Vehicles   79

AFV Miniatures   79
Weathering   80
Armored Conversions   82
Superdetailing Models   83
Combining Scratchbuilt and Kit Parts   83

7. Trains   91

The Railroad Empire   91
Materials for Assembling Railroad Models   92
Learning from Kits   92
A Caboose Kit Conversion   93
Etched Brass Kits   96
A Cab Conversion   99
Building a Steam Locomotive   100
Researching Model Locomotives   101
A Handmade Brass Model   106

8. Buildings   107

Cardboard Models   107
Building a Cutout Station   110
Building with Wood   112
Simulating Brick and Stone   116
Plastic Kit Conversions   117
Painting Tips   119

PART II
Building Materials and Techniques

9. Plastics   123

Precision Work   123
Surface Textures   126
Cementing Plastic   127
Embossed Details   129
Hole-Punching Techniques   131
Heat-Formed Plastics   132
Vacuum-Formed Plastics   132
Using Filler Putty   133

10. Wood   135

Making Master Patterns   135
Carving Shortcuts   136
Supersmooth Finishes   136
Fiberglass Finishes   138
Lightweight Wood   139
Duplicating Wooden Structures   140
Learning from Kits   141
Staining Wood   142

11. Metal   144

Brass and Nickel Silver   145
Cutting Sheet Stock   145
Etching Sheet Metal   147
Machining Metal   151
Milling Machines   151
Assembling Metal Parts   152
Soldering   155

12. Castings   159

Casting Materials   159
Making Molds   161
Casting with Epoxy   164
Using Other Casting Materials   165
Casting Inserts   166
Assembling Castings   167

13. Working Models   169

Model Engineering   169
Scale Power   172
Control Systems   173

Sources of Supply   177

Model Plans   177
Building Materials   179
Tools for Modelers   179
Clubs and Organizations   180

Index   181

Chapter 1
Model Building

E_{ACH} of us wants some particular material object that we cannot possess for one reason or another. Modeling can often place these objects within reach and provide the pride of achievement along the way. You can build it yourself and, if done well, your miniature will look so much like the real object of your dreams that it can fool a camera. You'll discover that the amount of effort needed to create a miniature replica does not always have a direct bearing on the realistic appearance or performance of the finished model. Some snap-together or almost-finished models can be assembled and painted in an evening or two to look and "act" exactly like the real thing. You may prefer to spend many months hand-shaping and fitting the parts from more complex kits or creating parts from raw chunks of wood, plastic, metal, or cardboard. There are some model builders who prefer to see their miniatures in action, rather than spend time building them; others enjoy the building more than the action. The model building hobby encompasses both extremes, and everything between. Somewhere between the snap-together kit and the scratchbuilt brass locomotive is the special place for you to develop your leisure-time skills. In this book, there is an affordable and attainable way to fulfill your dreams.

Scratchbuilding

Model builders generally refer to a miniature made mostly from raw materials rather than from a kit of preshaped parts as a "scratchbuilt" model. The strips and blocks of wood, metal, or plastic, the pieces of wire or rod, and screws or bolts are the raw materials. A modeler who assembles or creates a model from such materials is really starting from nothing or from "scratch." In some cases, the raw materials themselves are called scratch. These materials are cut or carved and shaped to match the contours of the full-size prototype that the model is supposed to duplicate.

Many of the early model kits of the thirties and forties were nothing more than blocks of wood with scale plans. The modeler was expected to use the plans in the so-called kit to carve the blocks of wood into the shape of the aircraft or automobile depicted in the plans. Some of the more expensive kits might include a few preshaped metal castings for the wheels or the aircraft's propeller, but most kits just included strips of wood to be carved into the proper shapes. There are still a few kits like this on the market, but over 99 percent of the wood, metal, plastic, or cardboard kits contain precut parts. Most kits include preshaped parts that need only be trimmed and fitted slightly before they are assembled and painted to complete the model.

There is certainly more satisfaction and pleasure to be gained from building a model from scratch than from a snap-together plastic kit. It's also true that it takes far more skill and years of practice to be able to build a model from scratch that is even close to the level of

Fig. 1-1 Wills Finecast offers a series of all-metal classic car models in 1/24 scale, including this 1933 M.G. K3.

detail of a plastic kit. The question, then, is whether it is worth the effort to learn to build a model from scratch. Fortunately, the skills you must acquire to build from scratch are best learned by building kits. That is really what this book is all about: acquiring the skills of building from scratch through the assembly of kits. The kits you assemble can also help you to decide just how much of your future models should be built from raw materials or "scratch" and how much from preshaped parts. Today's modelers, even the professionals, seldom build every detail of every model from raw materials. Parts from kits such as wheels or windows can be used to provide the details on what is otherwise a scratch-built miniature.

Kit Conversions

Modelers of the past were forced to construct every piece of every model from raw materials because there was nothing else available. Even simple items such as jeweler-size screws and nuts were difficult to find. A miniature replica of a steam locomotive assembled over a Lionel or American Flyer toy train chassis might be made from tin cans, old broom handles, and the cardboard that laundries

Fig. 1-2 James Newcomer used the doors from two AMT 1939 Chevrolet two-door sedans to make this 1/25 scale four-door sedan.

used to make shirts stiff. The rivets for the model might well have come from the pins used to hold that same shirt to the cardboard. All these primitive materials are still available, but today's model builder has an almost infinite choice of special materials and small detail parts to choose from to create a miniature.

The availability of such detail parts, along with the high standards of detail even on inexpensive "toy" models, has elevated the standards of all of today's models. A model railroader expects more detail on a $2 ready-to-run "toy" boxcar, for instance, than a custom builder of the forties would have included on a $20 handmade miniature. These inexpensive toys are both a blessing and a curse to the model builder: a blessing because the parts from the toys (and similar specially molded detail parts designed for hobbyists) can be used on scratch-built miniatures; a curse because the shirt cardboard boxcar is no longer going to be a very satisfying model when compared with "toy" train cars.

You have a right to expect this book to be a kind of course of instruction in how to build models. That is precisely what I would like it to be, but you may find that the "lessons" are not exactly what you expected. First, you must understand that it is often futile to try to build something better than what you can buy in a kit. Few modelers find reward in spending several hundred hours creating a miniature that looks, at best, exactly like what could be assembled from a plastic kit in a hundredth of that time. If there is some particular model you wish to build, you would be wise to make a thorough search to determine whether or not a kit is available to build it from. If you fail to find a kit, try to find a similar model kit that is available which you can modify to build exactly what you want. If, for example, you want a model of an Airabonita World War II fighter and you can only find a kit for the Airacobra, you'll discover you can easily modify the Airacobra to make an exact replica of the Airabonita. The work of adapting one model to duplicate a different prototype is called a "conversion" by aircraft modelers and a "kit-bash" or "cross-kit" by model railroaders. Other terms for modifying a kit (or for combining parts from two or more kits into a single model) include customizing, kit conversion, or kit customizing. The knack of combining two or more kits or of modifying the parts of a single kit is an intermediate step that most modelers take between assembling stock kits and building models from scratch.

Scale Models

One of the differences between a toy and a model is that the model is a precise replica of some full-size object in every dimension. A toy house, for instance, might have very small rooms and doors too low for the dolls that "live" there to enter. In a miniature of a home, however, every door and window and room will be in the same proportions as a full-size house. Truly accurate models even have such details as the window frames, the thickness of the walls, and the shape (at least) of the door handles and hinges reduced to match the balance of the structure. Miniatures that don't look "right" often have some mistake in the scale; the bricks might be ten times as large as they would be if they were reduced to the same scale as the rest of the house, the window frames might be too thick, or the pitch of the roof might be wrong. If every dimension

Fig. 1-3 Lloyd Jones used sheet plastic to shape the wings and fuselage for his 1/35 scale Loening OA-1A biplane.

of the full-size prototype is reduced by exactly the same amount for the miniature, that miniature is a true-to-scale model.

Some modelers go so far as to duplicate the invisible framing for the walls or to use exactly the same materials on the model that are used on the real thing. Such "pure" model building does not always result in a miniature that looks any better than one made using special materials suited for the particular model. An HO (1/87) scale house might have walls made from .060-inch (about 1/16-inch) thick Evergreen brand milled plastic sheets with plastic Grandt brand window castings, and look every bit as realistic as an identical model made from exact-scale strips of wood ("scale lumber"). Modelers often feel that the appearance of the finished model is far more important than what materials are used to build the model. Most modelers, in fact, find the hobby to be more enjoyable if the materials are selected for their ease of cutting and assembly to produce exact-scale dimensions.

"Scale" is another term used to state the proportions of the model in relation to the proportions of the prototype of the model. Scale is usually indicated as a fraction, so that a 1/87 scale model is 87 times smaller than its prototype. Model railroaders often use a system of letters that has been developed to identify the scale of their models over the past 50 years. Flying model aircraft kits (and plans) are sometimes described as an inches-to-the-foot scale. Table 1-1 lists most of the popular scales (proportions) used in model building. Most inches-to-the-foot scales appear on the triangular rulers used by draftsmen and sold in drafting supply stores. Rulers calibrated in the popular model railroad scale sizes (N, HO, S, and O) are available from shops that specialize in model railroad equipment. Any of these scales can be determined by reducing the measurement on the full-size prototype by the scale of the model. If, for example, the prototype's window is 9 inches wide, it would be 9 divided by 87 or 9/87 inches wide. This works out to .1034 inches. You can use a machinist's micrometer or a vernier caliper to measure the .1034 inches but it's a lot easier to buy an HO (1/87) scale ruler from a model

Fig. 1-4 Albert Hetzel's massive 1/48 scale Santa Fe Railway 2-10-2 was scratchbuilt from brass sheet and commercial castings.

Fig. 1-5 Dan Wilson used a kit's track and front wheels with a body scratchbuilt from sheet plastic to make this Japanese half-track.

railroad supply shop so that you can simply read the 9-inch markings from the ruler.

The popularity of a particular scale is the best possible reason for selecting that scale for your models. You'll be able to buy a scale ruler, or to use an architect's ruler, and there is a much better chance you'll be able to buy ready-made parts for your model. It would be foolish to build a miniature railroad locomotive to 1/100 scale, for instance, when 1/87 scale is so close to that size. The chapters dealing with each specific type of model will help you to decide which scale might be best for a model of that kind of real-world object.

Parts Building

By the time you have progressed through the stages of assembling stock kits and using kits for conversions, you may have developed your model building skills to the extent that you feel prepared to be a bit more creative by working from plans. You have a choice of either making every part of the model from raw materials for a truly scratch-built miniature or of using all the shortcuts you can find to locate parts that will fit your projected model.

The techniques presented in Chapters 9, 10, 11, and 12 provide the techniques you'll need to make just about any part imaginable from scratch. But, if you have been wise enough to select one of the popular scales for your model, you should be able to buy most of the smaller details. In some cases, you can find enough parts to collect a complete "kit." You may also find that you can work backwards from an existing kit by cutting up its major pieces and modifying its details to match the parts needed for your proposed miniature.

There really isn't a single model in this book that was built completely from scratch. The modelers who created the examples on these pages, however, did not want to spend the time making small parts when those parts were already available. The situation is similar to that of trying to scratchbuild a model when there is already a fine kit available; you can certainly do it, but why bother? If the available parts are at least a match for the ones you could make, then spend your time researching the prototype to be certain you have captured all of its essential shapes, details, and its subtle colors and shadings. The advantage of the detail castings, whether from kits or open-stock castings, is that even the perfectionist now has the time to add as much extra detail as he or she wants. It's finally possible for anyone with the patience to develop some basic skills to build a model realistic enough to fool anyone in a photograph.

Researching Model Prototypes

You might prefer to call the time spent researching your miniatures "watching trains" (or boats or cars or planes or whatever); either way, it's time that can be most helpful to you if you want to capture the "life" of the prototype. If you can visit the real thing, by all means do so. Take color photographs, make sketches with important dimensions, and generally get the feeling of the object just as an artist might for a painting or sculpture. You are, after all, creating a three-dimensional work of art; a sculpture that is supposed to capture not only the shape but the colors and even the textures of the real thing. You cannot do a creditable job working from either memory or photographs alone.

If you cannot visit the prototype, because

Fig. 1-6 A 101 Productions cut-out paper replica of the "Maxwell House" in Georgetown, Colorado, assembled by Sandy Lillie.

it's on the other side of the world or because it was demolished decades ago, then take a close look at similar objects as they appear today. Aged wood, riveted steel, flush-riveted aluminum, weathered paint, and polished metal look much the same today as they did 10 or 20 or 50 years ago (except for riveted aluminum, of course). There is little chance that every model you build can be based on a prototype that is easily accessible to you. Most miniatures you are likely to desire will probably be available to you only in photographs.

The object of your research into the "character" of the prototype for your model should be to find as many photographs of the prototype as possible. Most of the photographs you'll find will appear in books and magazines. You may be able to locate some sources of actual photographs in the advertisements in current magazines that deal with your favorite prototypes. There are often advertisements, for example, for color slides and color postcards of railroad equipment that may date back to the forties. Similar material is available for ships, structures, automobiles, aircraft, and rockets. You won't find many color photographs dating earlier than the forties because color photography was only perfected during the Second World War. However, there are plenty of color references available from historical societies and enthusiast groups for any period in history. You must do a considerable amount of reading to locate those sources. I've listed some of the suppliers of parts and materials and plans in the back of this book; their catalogs will get you started. Additional references are listed in the other Chilton books that deal with specific hobby areas. There are thousands of books on railroads along with additional thousands on ships, structures, automobiles, aircraft, and rockets. You must use these published photographs to verify any published plan you might be using.

The published plans depict the prototype at just one stage in its lifetime; almost anything you might wish to model has changed over its lifetime due to weather, repainting, modifications, or just plain wear and tear. Even the best plans should be used only as a guide to the sizes of the parts on your model. If you can locate enough photographs of the prototype, you may be able to make your own sketches using the published dimensions. If no dimensions are available, you should be able to judge the size of the prototype from known details such as door heights, tire sizes, or window sizes.

Take as much time as you need to obtain all the photographs and plans you need to com-

Fig. 1-7 The Imai/Scale Craft 1/350 scale "waterline" plastic kits can be displayed on that firm's plastic water display mat.

plete a detailed model; every experienced model builder has a tale of how he or she located a photograph or plan of a just-finished model too late to make a major needed correction. You can build a model that you'll know is accurate enough if you can locate at least a set of three-quarters front and three-quarters rear view photographs and a list of all principal dimensions of the prototype. Additional views and color references will certainly be helpful, and you may need many more photographs to capture the subtle shapes of some automobile bodies, ship hulls, or aircraft fuselages.

Working with Plans

One of the endless list of "Murphy's Laws" includes the maxim that the plans you want to use will be available in every scale but the one you need to build the model. There are two ways around this problem; convert every dimension on the existing plans to your scale as you build the model, or have the plans reduced or enlarged to your scale. Almost every town has a photostat shop to make copies of plans for the convenience of local architects, engineers, machinists, and contractors. The Yellow Pages list most photostat shops under the heading "Photo Copying." Unfortunately, that heading also includes photographers who will make copies of oil paintings, old photographs, and the like, so you may have to make several calls to find a firm that makes photostats.

The photostat shop will want a figure percentage for the enlargement or reduction of the drawing. There is a great chance for error in determining this percentage, so you'd be wise to persuade the shop to reduce a line on the original drawing that is, for example, 9 inches long to a length that will result in a plan of the scale you desire. This can serve as a double check for the math needed to determine the reduction.

The mathematics to reduce a plan from 1/48 scale to 1/72 scale are simple enough. Divide

Table 1-1 COMMON SCALE PROPORTIONS

Scale (Proportion of the Prototype)	Other Designations for Same Scale	Model Types
1/3800 (the smallest)	none	plastic rocket kits
1/1250	none	cast metal ships
1/1200	1 inch = 100 feet	cast metal ships
1/720	none	plastic ship kits
1/700	none	plastic ship kits
1/635	none	plastic rocket kits
1/600	½ inch = 50 feet	plastic ship kits
1/400	none	plastic ship kits
1/350	none	plastic ship kits
1/285	none	cast metal armor
1/250	none	cast metal ships
1/200	none	plastic ship kits
1/160	¾ inch = 10 feet N scale	N scale trains
1/150	none	plastic ship kits
1/148	2 mm = 1 foot	British N scale trains plastic ship kits
1/144	none	plastic aircraft and rocket kits
1/120	.1 inch = 1 foot TT scale	TT scale trains plastic ship kits
1/100	3 mm = 1 foot	plastic aircraft kits
1/96	⅛ inch = 1 foot	ship kits plastic aircraft kits
1/87	3.5 mm = 1 foot HO scale	HO scale trains plastic built-up armor models
1/76	4 mm = 1 foot 00 scale	plastic armor kits British 00 scale trains
1/72	1 inch = 6 feet	plastic armor kits plastic ship kits plastic aircraft kits
1/64	³⁄₁₆ inch = 1 foot S scale	S scale trains ship kits
1/48	¼ inch = 1 foot O scale	O scale trains ship kits plastic aircraft, armor, and rocket kits architectural scale
1/43	7 mm = 1 foot	cast metal cars and kits British O scale trains
1/36	none	ship kits
1/35	none	plastic armor kits
1/32	⅜ inch = 1 foot gauge No. 1	plastic aircraft and car kits ship kits No. 1 gauge trains architectural scale
1/28	none	plastic car kits
1/25	none	plastic car and armor kits
1/24	½ inch = 1 foot	plastic car and aircraft kits architectural scale
1/22.5	none	LGB brand outdoor trains
1/20	none	plastic car kits
1/16	¾ inch = 1 foot	plastic car and motorcycle kits outdoor trains (3½-inch gauge) architectural scale
1/12	1 inch = 1 foot	plastic car kits dollhouse miniatures outdoor trains architectural scale
1/10	none	plastic car kits
1/9	none	plastic motorcycle kits armor, car, and plastic motorcycle kits
1/8	1½ inches = 1 foot	outdoor trains (7½-inch gauge)
1/6	none	plastic motorcycle kits
1/4	3 inch = 1 foot	flying aircraft kits
1/3	4 inch = 1 foot	outdoor trains
1/1	12 inches = 1 foot/full scale	motion picture and TV sets and props

8

1/48 by 1/72 to arrive at the fraction 48/72 or .6667. If you wish to enlarge a plan in 1/72 scale to build a 1/48 scale model, do just the opposite: divide the scale of the plan by the scale of the model to arrive at the fraction 72/48 or 1.500. The rule is: scale of plan/scale of model. To double check your fraction, find or draw a 9-inch line, for example, on the 1/72 scale plan. On the reduced photostat copy of the plan this line should be .6667 times 9 inches, or 6.0003 inches. Make several such measured and mathematical reductions for a rough overall sketch of the width and length of the proposed model to be certain your proposed reduction (or enlargement) of the plan will produce the size plan desired. Photostatic copies vary considerably in price depending on the overall size of the original and the finished plan, but you can be certain the cost will be at least $10 and maybe as much as $100 for a very large photostatic copy. It's wise, therefore, to obtain two quotations for the work before you place an order for a photostatic copy.

When you are working from either an original plan or a photostatic copy, you must protect the artwork so that glue and cuts do not destroy it before the model is completed. The original can be protected with a sheet of wax paper, but I suggest you make some copies, so that the original plan or its photostatic copy will remain in perfect condition. If you are making a photostatic reduction or enlargement of a plan, you can often buy second or third copies for only a few dollars each. The best alternative for producing copies of either original plans or photostatic copies is to use one of the instant-copy machines that produces a paper copy in a few moments. These photo copies will often be slightly smaller than your original. There are two possible ways around this problem: use the paper copy as-is and ignore the small error in scale or try several different brands and models of copy machines until you find one that produces an exact-size copy. The small error that most copy machines make should not be a problem as along as you always refer to the copy and never to the original for any measurements. You can make several of these paper copies for a dollar and use them as templates to save time redrawing some plans. When you are working over a paper copy, using the paper copy as a gluing jig, protect the plan with a sheet of waxed paper so it doesn't stick to the model after the glue dries.

The plans published in books or magazines dealing with full-size aircraft or automobiles or ships or armor are not always the *only* drawings you need to make a miniature replica of that plan. Plans usually depict the machine as though you were looking directly at its side, top, front, or rear. If a panel on the machine angles toward you or away from you, that panel is going to appear foreshortened on the plan. If you intend to build a model directly over the plans, then you will have to redraw the plan so it can be used as a pattern.

The Smithsonian Institution offers plans for Lloyd Jones's Loening OA-1A seaplane (Figs. 3-4 through 3-8), but the wings on the aircraft angle upward (at a dihedral angle) so that they appear foreshortened on the plan. Lloyd's drawings in Fig. 3-4 show the wings laid flat. You can build a replica of the Loening OA-1A using only Lloyd's plans as your reference because those plans were created for the modeler.

The plans for the automobiles in Chapter 4 present the types of profile views where al-

Fig. 1-8 John Hathaway imports a series of precolored cutout paper models to build miniatures like this Mirage IIIC jet fighter. Photo by John Hathaway.

most every curved surface is foreshortened. In this case, however, you will want to carve the shape of the automobile from a solid block of plaster (Figs. 4-10 through 4-17) to make a plaster or a clear plastic replica, or you will want to carve the body from wood (Figs. 4-7, 4-8, and 10-1 through 10-7). With these techniques, all you want are the profile or "shadow" shapes of the side and the top of the body; in carving the rest of the model use your eye to determine if your model's curves match those of the prototype. Dan Wilson's 1/35 scale Japanese "Ho-Ha" half-track, in Chapter 6 (Figs. 6-6 through 6-14), provides an example of how you can determine the accurate length of any foreshortened part by referring to another view of that part. The true length of a wing with a dihedral angle, for instance, can be determined from the dead-on front or rear views of the aircraft.

Model-Building Tools

In the days of the block-of-wood "Kits," the only tools you needed to carve (build) a model were a penknife and a paint brush. You can build most of the modern snap-together plastic kits with the same tools. When your model-building skills improve, you'll want to try more challenging and complex kits, and, even if you use only plastic kits and parts, you will need to purchase several special tools. There is virtually no limit to the amount of money you can spend on tools. The tool I covet most is a jeweler's drill press with a feed sensitive enough that I can literally feel the drill at work. However, a new jeweler's drill press costs about $1600. Generally speaking, the harder the material you are shaping, the more expensive the tools will be to work that material. The amount of precision that the tool allows

Fig. 1-9 The basic tools for model building (*left to right*): steel rule, small hammer, X-Acto razor saw, self-clamping tweezers, pointed tweezers, flush-cut diagonal cutters, single-edge razor blades, hobby knife with blades, needlenose pliers, conventional pliers, Phillips screwdriver, conventional screwdriver, jeweler's screwdriver, medium-cut mill file, rectangular jeweler's file, round jeweler's file, emery board, and heavy scissors.

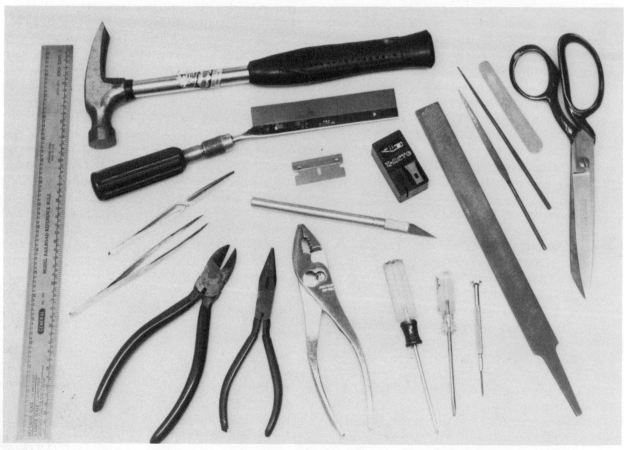

is another deciding factor in determining the cost of the tool, especially in the example of the jeweler's drill press.

There is really very little need for any electrically powered tools if you are working with thin sheet metal, sheet plastic, or sheet wood. Most of the tools that cut paper or cardboard will cut thinner sheets of wood, brass, or plastic; thicker sheets of wood or plastic can be easily cut with small saws and drilled with hand drills. You may want to use electrically powered tools to shape blocks or rods of plastic or wood, and such tools are almost essential for shaping thicker sheets of metal and metal blocks or rods. Electrically powered tools include items such as hand grinders, bench grinders, bench sanders, hand drills, bench drill presses, hand saws, jig saws, table saws, lathes, and mills. You'll find photos of most of the more useful power tools in the chapters that deal with working metals (Chapter 11) and wood (Chapter 10). Do remember that the same special tools (whether electrically powered or not) that are used to work thin sheets of plastic can often be used with thin sheets of wood, cardboard, or metal. Most of the tools and techniques described in Chapters 9, 10, and 11 can be used almost interchangeably with wood, metal, plastic, or even cardboard.

This basic assortment of small hand tools can be used to assemble or modify almost any plastic, wood, metal, or cardboard kit. Very few modern kits require drilling holes, so this assortment includes neither a pin vise nor an electric drill. The use of the hand-held modeler's drill or pin vise is described in Chapter 9 (as are punching and embossing techniques). The use of electric drills is described in Chapter 10 (along with tapping, turning, milling, and etching).

Virtually any model in this book could be built from scratch using just these hand tools and an electric drill with a drill press stand. The one "secret" to successful scratchbuilding is precise measurements and perfectly square openings and corners; therefore, I strongly suggest that you also invest in the surface plate, vise, and measuring tools described in Chapter 9 if you intend to build your models from raw materials. The same tools can be a big help in assembling many of the plastic, wood, cardboard, or metal kits that have many small parts and subassemblies.

Few of us can afford to purchase a complete set of model building tools at one time. If you are going to spread your tool purchases over a period of time, I recommend that you start with the following assortment of hand tools: the tweezers, X-Acto knife with number 11 blades, needlenose pliers, and scissors are probably the most important items in this initial package. I suggest you next budget $125 to $200 for a complete airbrush painting rig. A surface plate, vise, and accurate micrometers or vernier calipers should run to a total of less than another $100. Once you have these three major groups of tools on hand, you can consider a high-speed motor tool, electric saw, lathe, or some of the other powered tools. I have listed the purchases in an order that will allow you to (1) build most kits or kit conversions accurately, (2) paint any model so it looks as realistic as its basic shape, (3) have the necessary tools for accurate assembly of complex kits, kit conversions, parts-built, or scratch-built models. Any power tool except an electric drill is really a luxury that should be postponed until you have these first three tool categories in your shop. I have not mentioned the cost of a sturdy desk or workbench, or at least two sources of light and proper ventilation (a separate kitchen exhaust fan works well) for painting areas; you may have these essential "tools" available in your home. Remember that many cements, glues, solvents, and paints are highly flammable, so for your own safety your workbench and work area must be located well away from flames and be well ventilated.

Painting

There are at least a half dozen different brands of paints with special finely ground pigments and small-quantity bottles or cans that are sold for use with models. There are some tricks to applying these paints so they look like scale-thickness paint. I strongly recommend that you purchase an airbrush, but, in the interim, you can either apply paint with a brush or use the aerosol cans of spray paint for models. I do not suggest that you paint any model with gloss-finish paint from an aerosol can. The thinners and carriers needed for such paints apparently cause the paints to be relatively translucent so that very thick layers of paint, applied in six or more thin coats, are needed to completely cover the model. The bottles of gloss paints, from firms like Testors or Pactra, can be applied with either a brush or an airbrush to cover the model with just

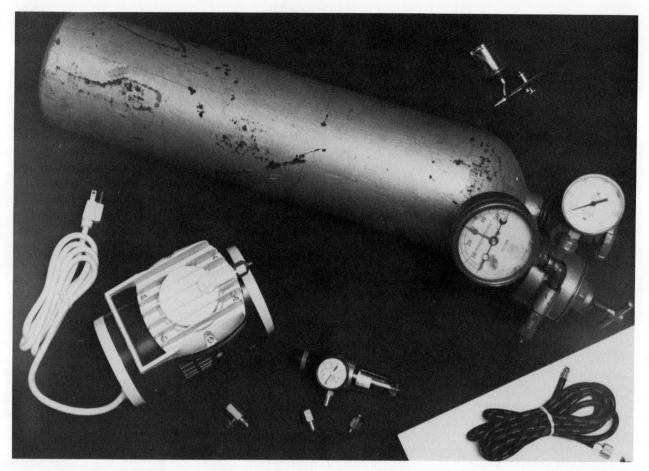

Fig. 1-10 An airbrush (*upper right*) can receive its air supply from a converted CO_2 cylinder (*center*) or from a small air compressor with a bleed valve and pressure gauge with a filter. A braided air hose (*lower right*) is necessary with either air supply.

one to three relatively thin coats. If you must use aerosol cans for glossy finishes, then use the flat-finish (nongloss) paints and, when they are dry, spray on a coat of Testors' Gloss-Cote from their aerosol can. The combination of a flat-finish coat of color and the clear gloss will give the best coverage of paint with the thinnest layer of paint possible without the use of an airbrush.

Selecting an Airbrush

The airbrush is a miniaturized version of the spray guns used to paint full-size automobiles. Badger, Wold, Binks, Thayer & Chandler, Paasche, and X-Acto make airbrushes that are generally available in hobby stores. Additional brands are sold by artists' supply stores. The airbrushes that offer external-mixing of the paint and air are only $10 to $20, but these inexpensive airbrushes have no adjustments for either air or paint flow, so the resulting spray pattern is about the same as what you would get from an aerosol can. These airbrushes do offer the advantages of allowing you to mix colors and to mix extra thinner with the paint like the more expensive airbrushes. Internal-mix airbrushes are available in single-action models and double-action models at prices ranging from $30 all the way to $100. Single-action airbrushes have a screw or knob that must be turned to adjust the flow of air through the brush; a push button controls the flow of paint. Double-action or dual-action airbrushes have a push button like the single-action airbrushes, but the button can also be moved back and forth to control the flow of air through the brush. When you become proficient in the use of the airbrush, you may want the advantage of the double-action brush so you can paint very small areas of a model and move immediately to larger areas without having to stop to adjust the air-flow screw or knob. Most airbrushes will spray a pattern as small as the periods on this page and a pattern as large as that from most aerosol cans.

The airbrush itself is the least expensive portion of any airbrush rig; the air supply is at least as important as the airbrush. It is possible to use an airbrush with aerosol cans of propellant or even an old automobile inner tube for the air source, but it is almost impossible to regulate the air pressure. A miniature air compressor will be needed to supply the air pressure for your airbrush. Today, a small compressor powered by an electric motor sells for between $50 and $150 depending on whether or not it has its own air tank and on the size of the electric motor. An air pressure relief or "bleeder" valve ($6 to $10) will be needed for a compressor that does not have its own air tank. The bleeder valve will help to smooth out the pulses from the compressor pump. You will also need an adjustable air-pressure valve with gauge ($25 to $50) and, if you live in a humid area, a moisture trap ($20 to $40) if these units are not already on the air compressor. Buy one of the braided air lines for your airbrush ($7 to $10) to replace the plastic line supplied with most airbrushes. I have an empty CO_2 tank that has been fitted with both capacity (air pressure) and line pressure gauges and valves by a welding supply shop. This air tank provides a silent supply of air which lasts about three years between fillings. Carbon dioxide is used, in place of "air," with this setup. The cost of the tank and gauges, however, is a bit more than most complete air compressor rigs.

Painting with an Airbrush

The only way to learn to use an airbrush is to try it again and again, with varying adjustments, until you get the general feeling of how it works. I suggest you adjust the air supply to about 25 pounds per square inch (psi) with the paint thinned with three-parts of the same brand of thinner, if possible. Lacquer thinner will work with most brands of paint except those with water bases; these can be thinned with water.

Hold the airbrush about 18 inches from the model and adjust the paint-flow screw or lever so the spray pattern is about the same as that from an aerosol can. When you are actually painting the model, always start the spray just to the left of the model, then sweep the spray pattern smoothly and evenly over the model. Do not let up on the paint-flow button until the pattern has passed off to the right of the model. You can move either the model or the airbrush. If you start the paint on the model, the pattern may create small splashes or spurts; similar spots can occur when you stop the spray. Incidentally, these same techniques apply when you are using aerosol cans.

The airbrush must be cleaned with thinner, sprayed through the brush, immediately after you are finished with each color. The aerosol can's nozzle can be cleaned by turning the can upside down and depressing the spray button so the gases in the can will blow the nozzle clean. If you are careful to clean the airbrush

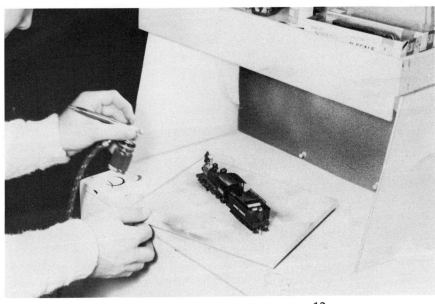

Fig. 1-11 Larry Larson constructed this roll-around spray booth using a furnace filter with a fan behind to catch the fumes.

immediately after spraying (so no paint dries inside its orifices), you should only need to disassemble it once or twice a year for a thorough cleaning. Be sure to wipe any paint from the inside of the cap and from the outside of the nozzle and the paint pickup tube with a rag dipped in thinner.

If the air supply is turned down to as low as 8 psi, the paint flow knob can be adjusted so that just a trace of paint will be sprayed from the airbrush's nozzle. You can then hold the nozzle as close as 3 inches from the model to produce 1/32-inch wide stripes or dots (with slightly fuzzy edges, of course). For really smooth gloss finishes, leave the air pressure setting at about 8 psi and open the air-flow screw or lever on the airbrush most of the way. Spray about 9 to 12 inches from the model to allow the paint to flow over the surface. Greater air pressure will result in more of the dusty or flat-finish effects needed for aging or weathering effects. My Chilton *Display Model Aircraft* book contains some additional information and tips that you may find useful for airbrushing any type of model.

Applying Decals

Decals for lettering, stripes, and other small surface markings are included in many kits. Thousands of other alternative decals are available for all types of miniatures from the various hobby shops and mail order firms. To apply these decals soak them in water until the glue that holds them to their paper backings dissolves; then apply the decal to the surface of the model. There are a few model railroad markings that are applied by simply rubbing the paper backing until the marking adheres to the surface. These markings are called "dry transfers" for obvious reasons. There is an almost unlimited selection of lettering styles and decorative trim available from the dry transfers sold for use by draftsmen and architects. These types of dry transfers are only sold by drafting or artists' supply stores.

It is almost impossible to apply either dry transfers or decals to the surface of a model that has been painted with flat-finish paint. The microscopic grain of the paint that makes it "flat" also prevents the decal or dry transfer from gripping the surface tightly. Small air bubbles will be trapped beneath the decal or dry transfer to make it appear cloudy. If your model is going to be decorated with decals it must be painted with glossy paints or the flat-finish paints must be covered with a light coat of clear gloss paint before the decals are applied.

Cut all around each decal about 1/16 inch from the printed portion using scissors. This will allow you to apply just one decal at a time. Hold the decal with tweezers and dip it into warm water for about 10 seconds; then place the decal and its paper backing on a paper towel or blotter. Let the water soak into the paper backing for about a minute to dissolve the glue. When the decal can be moved without moving the paper backing, it is ready to be applied to the model. Hold the decal and the paper backing, again, with tweezers while you position them on the model. Hold the decal itself with the tip of a hobby knife blade while you pull the paper backing from beneath it with the tweezers. Push the decal gently down on the surface of the model with a small paint brush dipped in water. If the surface beneath the decal has details like rivets or ribs or if the surface is curved, you must use a decal-softening fluid to allow the decal to adhere to the surface as tightly as the paint. Micro Scale's Micro-Set and Testors' Decal Set are mild softeners that can be applied to the surface before the decal is in place.

Micro Scale's Micro Sol, Walthers' Solvaset, or Champ's Decal Set are applied over the surface of the decal immediately after the decal is in place. The decal cannot be moved after it is coated because the softener is almost the consistency of paint. You may need to use as many as a dozen applications of the decal soft-

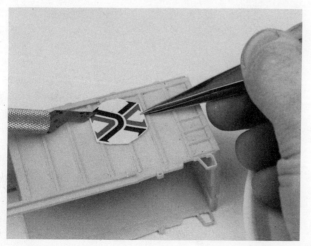

Fig. 1-12 Pull the decal's paper backing with tweezers to slide the paper from beneath the decal while you hold the decal in place with a hobby knife.

ening fluid to force the decal to snuggle tightly around some details. Allow the decal and the fluid to dry for at least 24 hours, then lightly scrub away any traces of residue from the model.

Let the model dry again overnight and then spray on a light coat of clear flat or clear gloss paint to produce the finish you desire. This final coat of clear paint will both protect the decal and match its gloss (or nongloss) to the balance of the model so the decal truly looks painted on. Be sure to try whatever clear paint you use on a scrap of decal; many brands of clear paint will dissolve or curl decals. I have not experienced any problems with Testors' aerosol cans of Glosscote or Dullcote, Pactra's clear gloss and clear flat, or the Micro Scale bottled clear flat and clear gloss, but you should test even these brands. Every brand of gloss I have tried does, however, attack or etch the styrene clear plastic windows in most plastic kits, so it should be applied before the windows are installed.

Fig. 1-13 The Jarmac drill press is not as precise as a true jeweler's-quality tool, but it will do nicely for modelers at about one-tenth the price. The Dremel speed control (*left*) or Jarmac's optional foot treadle control can be used to vary the speed of the drill: it provides slow speeds for drilling plastics and higher speeds for drilling metals.

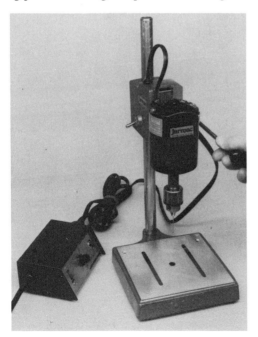

Warning to Enthusiasts

There's a pretty good chance you are already interested in at least one of the categories of models presented in this book or that you have some experience working with wood, plastic, metal, or cardboard. I almost regret the need to divide a book like this into chapters because those who feel they know what they like may skip over the chapters that do not, apparently, concern them and read only those whose subjects they know. Please, read every word in every chapter. Every segment of the hobby has something worthwhile to contribute to every other segment. Those who build plastic aircraft kits could teach model railroaders much about the art of airbrushing, for instance, while model rocket builders could teach all of us something about scrounging parts from other kits and outside sources for purposes far removed from their intended uses. The techniques used to impress rivets in sheet plastic or to make rivets with the dimple-and-paint-drop method work equally well on metal or cardboard models. The lathe-turning, milling, and shaping techniques needed to remove metal are virtually the same as those needed to shape blocks of wood or plastic. I could not repeat the same techniques over and over again for the chapters on working with wood, metal, plastics, or castings, nor could I repeat the techniques used by rocket modelers that are also applicable to aircraft, automobiles, ships, armored vehicles, structures, or railroads (or vice versa). You should be able to find at least one or two tips or techniques in every chapter of this book that can be used for your favorite types of miniatures made from the materials you like best. Don't let your enthusiasm for one type of miniature and one type of material allow you to overlook the lessons that have been developed over half a century of modeling.

I've discovered that the best professional model makers have one thing in common: each of them uses a wide variety of materials and techniques from every area of the hobby to create his or her patterns and prototypes for future kits. I've drawn on the experience of these professionals, as well as the expertise of some very clever amateur model builders, to show you how you can use kits to develop the model-building skills of the scratchbuilder.

PART I
Model–Building Projects

Chapter 2

Rockets and Spacecraft

*F*ANTASY is very much a part of every type of model building project; the use of one's imagination is just more obvious when applied to rockets and spacecraft than to a steam locomotive. The space-age segment of model building encompasses all the forms of the hobby that are included in the aircraft, automobile, ship, or railroad models. There are, for example, model rockets that are historical replicas of early craft like the German V-2, complete with camouflage and weathering. There are miniatures of the most modern rockets and, of course, replicas of the spacecraft proposed for the future and projected in the creations of science fiction films and television. Some of these rockets are designed to be nothing more than extremely realistic static or display models whereas others not only fly but separate into two or more stages, and some even take pictures during their flights. I have to include all types of fantasy models in the "rocket" category because there is really no other logical place for extraterrestrial vehicles like the NASA Moon Buggy, futuristic robots, and antigravity machines from fact and fantasy.

These machines, too, are the subjects of both static and motorized miniatures including radio-control robots and hovercraft that appear to perform much like the fabled antigravity machines. The hovercraft is very similar to a helicopter with the body built around the rotating blades or fan, except that the hovercraft has a skirt that almost touches the ground. Hovercraft are in use in swamps and as ferry boats over tumultuous bodies of water such as the English Channel. A model of a hovercraft can be made to skim just an inch or two above the ground as though it really were an antigravity device from the future. You'll need the skills of an experienced flying model aircraft modeler to be able to scratch-build a hovercraft; the true core of the rocket modeling hobby, however, consists of somewhat swifter craft.

The Perfect Fantasy

Model rockets and spacecraft encompass both the simple and the complicated extremes of modeling. There are dozens of very accurate cast-metal prepainted miniatures that are intended for the toy market but with enough details and precision to qualify as models. There are also some very simple snap-together plastic kits and some wood, plastic, and cardboard flying model rockets that can be assembled and painted in less than an hour. There are also very complex plastic rocket kits and multistage wood, plastic, and cardboard flying rockets.

The most challenging aspect of the model rocketry hobby, however, faces those who would design their own rockets. The trick with a display model rocket is to create a futuristic design that comes close to the credibility and realism of those depicted in the motion pictures and television epics. The film makers have an advantage, of course, in that they can

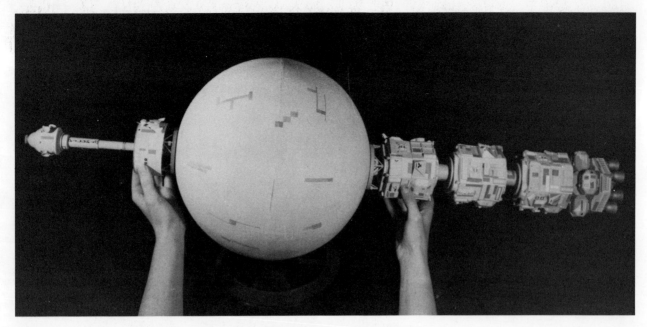

Fig. 2-1 The key ingredient in any static model of a spacecraft is ingenuity; that's a painted globe in Kevin Atkin's model. Photo courtesy of Kevin Atkins.

use lighting and special effects to make their miniatures appear to fly. The modeler can substitute imagination for the special effects, but only if his or her model looks like the real thing—failure to achieve this almost always results in a model that looks like it came from a third-grade science class. I don't want to present insurmountable obstacles to your pastime (or future career) as a spacecraft designer, but you should be aware that the creation of your own spacecraft design is only going to be successful if you have some experience in building kits that duplicate the designs of others.

The model kit manufacturers who produce plastic display model kits and those who produce flying model rockets are always aware of trends in fantasy vehicles. You can expect to find kits for both static and flying models of whatever space fantasy film or television series is currently popular, on the shelves of most hobby shops. These models can be assembled, painted, marked with decals, and even weathered just as though they were World War II aircraft. The sources of information on the prototype, however, will have to be taken from the science fiction and fantasy magazines that show color shots of these space vehicles from motion pictures and television.

The model kit manufacturers seem to make the same kind of mistakes they make with aircraft kits: the cockpit canopies are generally too thick, the interiors often lack details or may be out of scale (not proportioned to match the balance of the model), and the small antennae and other fine details are molded thicker than they should be. It's a challenge to the abilities of any model builder to make one of the kits look as real as the miniatures used for films; the film-makers' models are almost always four to ten times as large as the models in the hobby kits.

Once you've perfected the art of assembling out-of-the-box kits, the next step will be the addition of extra details and correction of any mistakes on the kit. From there, you may want to go a bit further to assemble your own creations. However, even the professional model makers who create the space vehicles for television and motion pictures use detail parts and pieces from the plastic models you'll find on your dealer's shelves. These professionals are trained to be clever so you'll see things like tank treads, automobile engine blocks, truck wheels, and other details serving to simulate windows, vents, antennae, and other outer space paraphernalia.

Do-It-Yourself Designs

The use of detail parts from plastic model kits on a special-design space craft is only a shortcut in the detailing of the final model. The real challenge in building such models comes from the overall shape and size of the

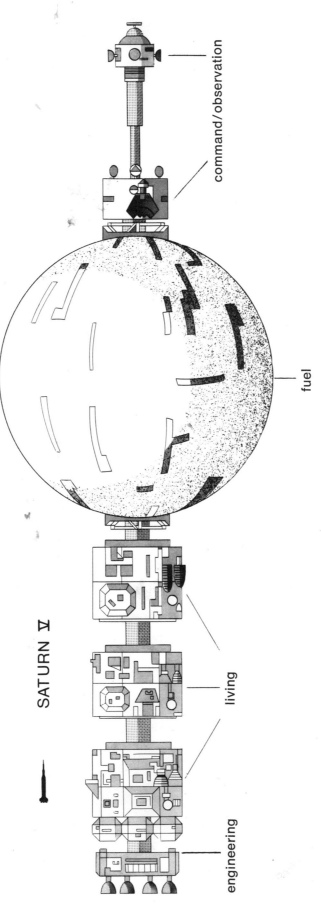

Fig. 2-2 The Enzmann Starship is a hypothetical spacecraft that would use the round chamber to house its fuel supply. Courtesy of Gates Planetarium; copyright © 1980, Denver Museum of Natural History.

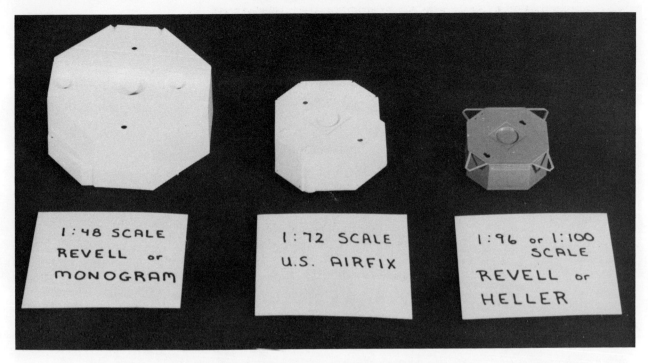

Fig. 2-3 The three octagonal chambers that simulate the living quarters can be made from any of these "Lunar Module" kits. Photo courtesy of Kevin Atkins.

space vehicle. You should be able to fabricate simple sheet plastic boxes of all shapes using the techniques described in Chapter 9. It's also helpful to know about the various H beams, channels, domes, tubes, and rods that are available in plastic from firms such as Plastruct, in wood from Northeastern, and in metal from Special Shapes, K & S, and Milled Shapes. You should also keep your eye open for everyday items that might serve as the basis for a space ship. Plastic soup plates really can look like flying saucers if you find the right shapes, metal and plastic cooking wear can sometimes be stacked, mouth-to-mouth, to form chambers for a space vehicle, and don't forget the inexpensive paper globes painted to look like planet earth or the moon. The ball or globe shape has inherent design strengths that make it popular with many proposed spacecraft designs for soon-to-come real-life explorations in space. Triangles, hexagons, polyhedrons, and other self-supporting shapes are also popular bases for both real and fantasy space vehicle designs. These flat-surfaced shapes can be fabricated easily from sheet plastic. There are hundreds of books that illustrate spacecraft that are actually in the planning stages and others that are merely products of the artist's imagination. I would suggest that you use one of these existing designs for your first attempt at a do-it-yourself spacecraft model. When you understand how to adapt sheet plastic, structural shapes, and odds and ends from other kits to duplicate an existing design, then you'll be ready to create your own space vehicles.

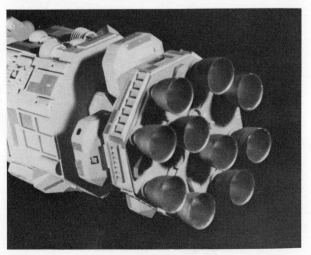

Fig. 2-4 You will need to learn to improvise when creating your own spacecraft designs. The exhaust nozzles on this miniature are from a "Space 1999" kit, but you can substitute anything with a similar shape, including jet aircraft exhausts or even jet aircraft nose cones. Photo courtesy of Kevin Atkins.

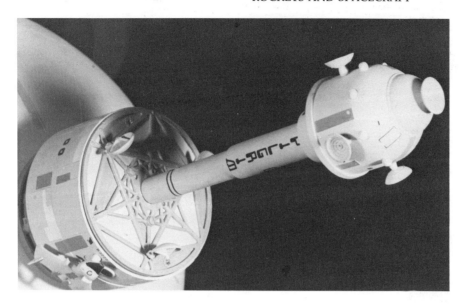

Fig. 2-5 Most of the parts on the nose section of the Enzmann Starship model can be fabricated from Plastruct tube and sheet. Photo courtesy of Kevin Atkins.

Assembling Plastic Kits

The highly detailed parts that you'll find inside any plastic model kit are deceiving; they appear to be an almost-finished model right in the box. Those parts must always be carefully trimmed, fitted, cemented together, filled with putty, painted, marked with decals, and weathered to look as nice as the artwork on the box lid. There is no such thing as a shortcut around any of those steps, even with the pre-colored, snap-together kits. The only possible way to build a miniature that looks like it is, was, or could be from real life is to take the time to do everything right. Because a model is only a fraction of the size of the real thing, flaws such as gaps between the parts, unpainted plastic surfaces, crooked decals, and the like appear many times more obvious than such flaws would appear on the full-size object. Remind yourself, if you must, that model building is a precise and perfect hobby. It takes at least as much patience as skill to assemble the simple plastic kits.

The kit-assembly process begins when you open the box. Do not break the parts from their molding sprues or trees just to see how the model will look when it is assembled. Each of those parts must be carefully cut from its sprue or tree with a sharp hobby knife; if you try to break off the part, you'll almost always leave part of the detail or the part's edge behind. The instruction sheets are always helpful in determining just which shapes are details, which are aligning pins, and which are simply excess plastic that has oozed out of the edges of the mold as "flash." Use the instructions as a guide to carefully trim each part from the molding tree so the parts can be test-fitted before any cement is applied.

You will need to use both bottled (liquid) cement for plastics and tube-type (thick) cement for plastics to assemble most plastic kits. The tube-type cements should be used as just pin-size beads on the longer joining edges of fuselages and wing edges. If you apply too much of the tube-type cement, the solvents may cause the plastic to sink, and this sinking can continue for a year or more! The tube-type cements dry a bit slower to give you time to spread the cement and get the parts together. Do not attempt to use the cements as a filler for seams or joints that do not fit tightly. Improperly fitted joints should either be filed to fit or the seam left open so it can be filled later with the automobile body "spot" putties. A single drop of the liquid cement should be enough for most smaller joints. Assemble the parts immediately after touching the joining surfaces with the drop of liquid cement. The liquid will flow along the seams between the parts by capillary action. Be careful, though, because the liquid cement will also find its way between your fingers and the model.

If the parts are warped enough so that clamps must be used to hold them together while the cement dries, then be extremely careful to avoid touching the actual seams with the clothespins, hairpins, or rubber bands that you are using for clamps. A pair of wooden toothpicks, placed along either side of glue seams, will keep rubber bands or other clamping devices from coming into contact with the liquid cement. Allow the cement at least 48 hours to dry before you apply any automobile body filler to the seams or before painting the model. If body filler must be used, it can be sanded or filed (with jeweler's fine-tooth files) flush with the model's surfaces within about 6 hours after the filler is applied to the model.

White glue (intended for use on wood) is best for installing the clear windows in any plastic model; the solvents in any plastic cement will fog or etch the clear plastic. Five-minute epoxy or Goldberg's Super Jet cyanoacrylate cement are best for tiny high-strength joints like aircraft landing gear, aerials, guns, or bomb racks.

Learning from Kits

There are several methods used by the plastic kit manufacturers to provide details and assembly alignment for their kits. You can learn a great deal about assembling and modifying plastic kits if you purchase an assortment of at least a half dozen different kits in different scales from at least three manufacturers. There's no need to assemble all of the kits, but you should go at least as far as the test-fitting stage so you thoroughly understand how the different models are supposed to go together. Complete at least three of them, using the liquid and tube-type plastic cement and body filler to completely hide the seams between the plastic parts. Paint these models and apply decals as described in Chapter 1.

The parts from the kits you have not glued together may be used as components for a custom-design space vehicle that you create using your own imagination. You will have to draw on the experience gained from assembling those first three or more kits, however, to develop some understanding about the design and assembly of plastic models in general. The process of combining parts from two or more kits to create a "conversion" is really the same as assembling a stock single kit except that you must do some sawing, filing, and careful fitting to mate some of the parts from the second kit to those of the first kit. The addition of plastic sheet parts, plastic shapes, and the use of other materials also requires cutting and fitting. Do try to use all plastic parts for your first "conversion" models; brass or cardboard or wood parts can be cemented to plastic using either 5-minute epoxy or Goldberg's Super Jet, but those multimaterial models can require extra precision in fitting the parts together.

The Enzmann Starship Model

Kevin Atkins's miniature of the Enzmann Starship was created for use in a planetarium show about interstellar flight. You can duplicate Kevin's model using parts from plastic Lunar Module kits, Plastruct tubing, leftover sprues from other kits, and one of the cardboard globes. The model in the photographs (Figs. 2-1 and 2-5) was assembled using three 1/48 scale Revell Lunar Module kits with a 13-inch-diameter Earth globe. Slightly smaller globes will be needed if you decide to use the Airfix 1/72 scale Lunar Module kits or the 1/96 Revell or 1/100 scale Heller Lunar Module kits. The nose section on Kevin's Enzmann Starship is a modified nose from Lindberg's "Star Probe Space Shuttle." If you elect to use smaller Lunar Module kits than Kevin's model, you may have to fabricate your own observation nose from parts of other spacecraft kits. Kevin also used four of the octagonal pods from the Eagle spacecraft from the British Airfix "Space 1999" series, but those kits are rather difficult to locate in America. The exhaust nozzles are also from this same kit.

Don't be discouraged if you cannot locate exactly the same parts used for this model; the fundamental principle of modeling spacecraft of the future of fantasy vehicles is imagination. There are dozens of other plastic space vehicle kits that can provide the various chambers of the Enzmann Starship since there really is no such vehicle. You may even want to assemble an entire fleet of space vehicles based on the globe-shaped fuel cell concept. The larger vessels, for example, could be made up of a triangular group of three fuel globes.

Making Rockets that Fly

Spacecraft model building is really divided into two categories: models that are intended strictly for display purposes and those that really do fly under rocket power. Both segments of the hobby include miniatures of existing and proposed rockets as well as fantasy creations from the possible distant future. There is no limit to the design, shape, or size of display model rockets other than that they fit in your room. The flying model rockets, however, must actually contend with both gravity and air flow because their flights are never more than 2500 feet. The models reach that altitude at true rocket velocities; the rockets reach their maximum altitude in a matter of seconds. The replaceable rocket engines are designed so they can force the ejection of a parachute at the apogee (peak) of the flight. Some designs eject the engine alone so that the rocket can glide back to earth; others have multiple gliders built into the design.

The rocket engines are solid-propellant rockets encased in a thick cardboard tube with a ceramic nozzle and clay endcap. A new engine is used for every flight. There are more than two dozen standard rocket engines with a variety of thrust, thrust-time, and delay times (for ejecting parachutes or gliders). Estes, Centuri, and Flight Systems are the leading manufacturers of both flying rocket kits and model rocket engines. With the exception of some plastic nose cones in a few kits and one-piece fin assemblies in a few beginner kits, all of the flying model rockets have balsa wood nose cones, balsa sheet wood fins, and bodies of special cardboard mailing tubes. These materials are actually far lighter than plastic and much safer. I do not recommend that you attempt to construct any type of *flying* model rocket from plastic.

The best way to learn about the hobby of flying model rockets is to purchase one of the Estes or Centuri Flying Rocket Starter Outfits. These kits contain everything you need from the rocket kit itself to the launch stand and engines for several flights. There is a choice of rockets in these starter kits that can range from a simple precolored kit that can be assembled in minutes (plus time for the glue to dry) to rather complex replicas of some of the science fiction/fantasy rockets.

Most of the kits assemble just like the "Double Trouble" rocket described in this chapter except that some of them have precut fins. The "Double Trouble" is a two-stage rocket that is powered by two separate rocket engines. When the first engine (the one nearest the

Fig. 2-6 Estes Industries' Designer's Special kit includes enough wood and cardboard and plastic parts to build five or more rockets.

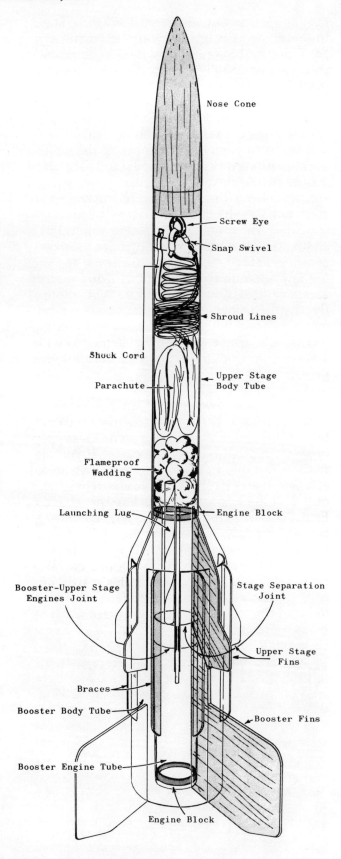

Fig. 2-7 The "Double Trouble" is a proposed rocket kit design that is now a collector's item; it never made it into a kit. This cutaway view identifies the parts of this flying rocket model. Courtesy of Estes Industries, Inc.

ground) burns out, it fires a forward charge to simultaneously ignite the upper engine and eject the first stage (also called a "booster" stage). The ultralight booster stage simply tumbles to the ground. The upper stage then proceeds upward until its engine burns out and, after a short period of upward coasting, fires a foward charge to eject the nose cone and release a parachute from inside the rocket body (see cutaway drawing Fig. 2-7). The parachute is attached to the nose cone with a string (shroud line) and to the rocket body with a rubber-band "shock cord." The nose cone and upper stage float to the ground beneath the parachute. The beginner's rockets are single-stage rockets to give you a chance to try different rocket engines without the complications of two stages. Don't attempt to fly any two-stage rocket until you have become familiar with the performance characteristics of a single-stage rocket launched with a variety of different thrust and duration engines.

Building a Two-Stage Flying Rocket

Most dealers carry assortments of rocket body tubes and nose cones as well as a supply of sheet balsa wood for nose cones. If you are really interested in designing your own rockets, however, I suggest that you purchase one of the Designer's Special kits like that sold by Estes. The kit includes nose cones, body tubes, two-stage adaptors, parachutes, and a design manual and it costs about half what the total price of all the components would be. The parts list (Table 2-1) includes Estes' part numbers on the individual components we used to build the model in the photographs. All of them were included in the No. 1463 Estes' Designer's Special kit (Fig. 2-6).

Cut out the two fins and the brace patterns (Fig. 2-9) and trace them onto two sheets of 1/16-inch balsa exactly as shown in Fig. 2-12 so that the grain will run in the proper direction. Use a steel ruler to mark the straight edges and to guide your hobby knife blade when you cut the parts from the balsa. Make three or four light cuts rather than one heavy one. Hold the end of the steel ruler against a scrap of 1

Table 2-1
Parts List for "Double Trouble" Rocket

Quantity	Estes Part Number	Description
1	BNC-20B	nose cone
1	BT-20	body tube .736 × 18 inches
1	BT-60	body tube 1.637 × 18 inches
2	EB-20A	engine blocks (3 per package)
2	BFS-20	1/16 inch thick balsa fin stock, 3 × 9 inches
1	2321	launching lug (12 per package)
1	2281	screw eye (6 per package)
1	2276	shock cord
1	2292	snap swivel (12 per package)
1	2264 (PK-12)	12 inch parachute kit

Note: Full packages of the smaller parts must be ordered from your dealer even though only one or two parts may be needed.

× 4 lumber so you can rotate the body tubes while you mark their cutoff lines (Fig. 2-13). Use a sharp hobby knife to cut the body tubes to length along the lines just marked. The "engine block" is a thin band of paper that is glued inside the body tube (see cutaway Fig. 2-7) to retain the rocket engines. Cement the one for the booster rocket 1/16 inch from the end of the 3 5/16-inch-long by .736-inch-diameter BT-20 booster tube (Fig. 2-8).

The upper stage body is an 11-inch-long piece of the same size tube. Cement the engine block for the upper stage 2¼ inches from the end of the tube (Fig. 2-9) if you are using Series I rocket engines. This will allow the engine itself to project about ½ inch from the bottom of the upper stage to help join the upper stage to the lower stage (booster). That ½ inch of engine will actually fit inside the upper half of the booster rocket's inner body. Most modelers prefer white glue to install the engine blocks. I used Super Jet cyanoacrylate to install the fins, but this cement will glue your fingers together so its difficult to use *inside* the body tubes.

Cut one of the BT-60 body tubes (1.637 outside diameter) to a 5-inch length to make the outer tube for the booster rocket. Tape the booster body cutting guide (Fig. 2-11) around

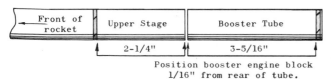

Fig. 2-8 The thin cardboard rings called "booster engine blocks" must be positioned inside the body tubes as shown. Courtesy of Estes Industries, Inc.

MODEL-BUILDING PROJECTS

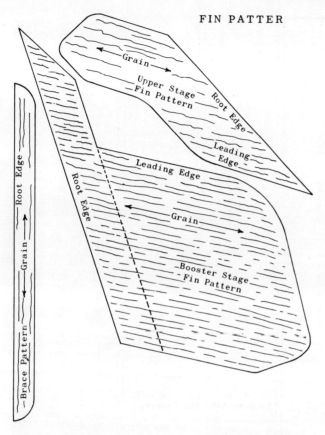

Fig. 2-9 A full-size template with grain pattern direction for the 1/16-inch balsa wood parts for "Double Trouble." Courtesy of Estes Industries, Inc.

this tube and carefully cut out the "receiver slots" for the fins on the upper stage rocket and for the fins on the booster stage rocket using a hobby knife with a new blade. You should have the tube with six slots on one end and three slots on the other when you are through cutting. Use the dotted lines to mark the positions of the braces on the tube.

Stick a straight pin through the pattern and into the tube at several places along those dotted lines to mark the fin positions on the tube before you remove the paper pattern. The fin spacing guide (Fig. 2-11) is needed to align the 1/16-inch balsa wood fins on both the upper-stage and lower-stage body. The six fins are to be cemented to the upper stage body *only*, using the six arrows on the fin spacing guide to locate the fins. I suggest that you mark the upper and lower-stage body tubes, using the spacing guide, and then assemble the upper

and lower stages temporarily using a burned-out rocket engine to keep the two tubes temporarily aligned. Slide the lower (booster) stage's three fins into the slots in the 5-inch-long tube and use *both* the marks you made on the smaller diameter, 3 5/16-inch, long tube to align the fins. Glue these three fins to the sides of both the 5-inch large-diameter tube and the marks on the 3 5/16-inch diameter tube. Sight down the tube to be certain the fins are perfectly vertical where they touch the tubes before the glue dries.

The six fins for the upper-stage (11-inch long) tube can now be positioned into the six notches you cut into the top of the 5-inch booster tube (see Fig. 2-14) and aligned with the six marks you made on the smaller diameter tube. Again, sight down the tube to be certain the fins are perfectly vertical in relation to the inner and outer tubes. These six fins must be glued *only* to the 11-inch-long upper-stage body tube. Be careful that no glue accidentally flows into the spent rocket engine that is joining the two small-diameter upper and lower engine tubes.

The six 1/16-inch balsa wood braces can now be cemented to the marks on the large-diameter tube. I've painted the entire booster rocket stage a darker color (red) and the upper stage a lighter color (yellow) so you can see where one stage begins and the other ends. The six fins on the upper stage *must* be perfectly free to slide out of the slots in that large-diameter tube or the two stages will not separate. If that happens, the upper-stage engine will destroy the lower-stage rocket! Shave the notches with a sharp hobby knife or sand them with fine-grade sandpaper, if necessary, to get a fairly loose fit over those upper-stage fins. Check the fins again, after the model is painted, to be certain that the paint has not made the fins thicker (or the notches narrower) and sand the notches if there is any sign of a tight fit. Glue the 2 3/8-inch-long "launching lug" (it looks like a piece of a paper drinking straw) to the side of the upper body as shown in Fig. 2-7.

Thread (rotate clockwise) the metal screw eye into the flat end of the balsa wood nose cone (No. BNC-20B). Remove the screw eye and dip its pointed end into some white glue,

Fig. 2-10 Wrap this full-size fin-spacing guide around the body tubes to locate the fins for the booster and upper-stage rockets. Courtesy of Estes Industries, Inc.

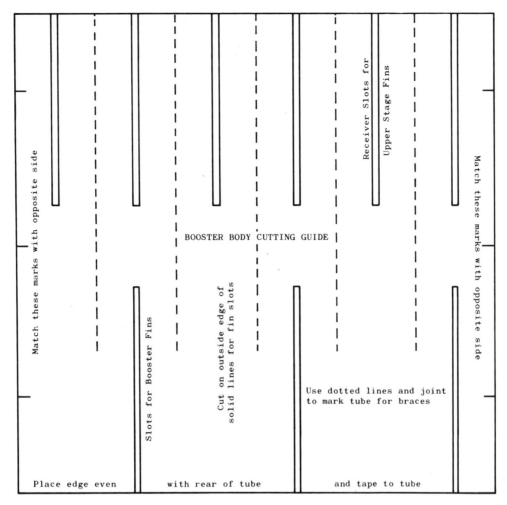

Fig. 2-11 Use this full-size template to carefully cut the notches in the booster's larger diameter body tube. Courtesy of Estes Industries, Inc.

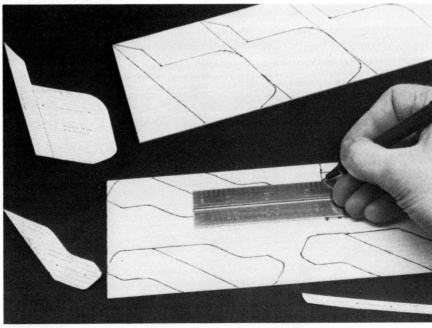

Fig. 2-12 Use a steel ruler to mark the fins and to guide the hobby knife when you cut them from the 1/16-inch balsa wood sheets.

Fig. 2-13 Rotate the body tubes against a scrap of 1 × 4 lumber while you hold a marking pen against a ruler to mark the cut lines.

and then screw it back into the nose cone. Cut two 3/8-inch long slits around the diameter of the upper body. Cut one of the slits about 3/4 inch from the top of the upper body tube and the second slit about 3/4 inch from the top (see Fig. 2-7). Push the area between the slits inward enough so that the rubber-band shock cord can be threaded through the piece of cardboard tube. Push the notch back flush with the tube to trap the end of the shock cord and glue the slits and the shock cord together with white glue. Tie the free end of the shock cord to the screw eye.

Assemble the parachute and its shroud lines according to the instructions furnished with the parachute and attach the shroud lines to the metal snap swivel. Hook the swivel onto the screw eye. The rocket engines for the first flight (to check the stability of the rocket) should be a 1/4A.8-0 for the booster engine and a 1/4A.8-4 for the upper stage.

Painting a Rocket

The relatively coarse grain of the balsa wood will be accentuated by almost any kind of paint. The grain can be filled with several coats of sanding sealer liquid that is smoothed, after it dries overnight, with fine-grade sandpaper. The sanding sealer is available at shops that sell flying model aircraft supplies under the names "sanding sealer" or "balsa fillercoat."

There are two schools of thought on smooth finishes for both flying aircraft and flying rockets; one theory suggests that the air will flow more easily over a supersmooth surface like that achieved with several layers of sanding sealer and two or more coats of paint (sanded smooth between each coat with No. 600 wet-or-dry sandpaper dipped frequently in water). The alternative theory is that the slightly rough finish of unfilled but painted wood might actually set up a boundary layer of air that will serve as a kind of lubricant for the surface's push through the air. The sanding

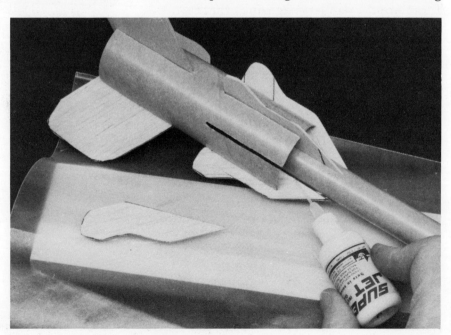

Fig. 2-14 Support the model with waxed paper to catch any dribbles of the Super Jet instant cement or white glue.

sealer and any extra coats of paint will add a noticeable amount of weight to a rocket model and that, too, can impair its performance. You can use either enamel or the special fuel-proof paints for flying models. Never attempt to apply fuel-proof paints (or lacquers) *over* any enamel or acrylic paint because the first coats of paint will be dissolved and wrinkled by the more volatile fuel-proof or lacquer paints.

If you are going to mark the surface of your model with decals, you must use many coats of decal-softening fluid to force the decals to conform to the shape of the wood grain. The decals should also be protected with a spray-on coat of clear paint such as Testors' Dullcote or Pactra's clear gloss.

Researching Rockets and Spacecraft

The books and magazines that deal with actual and science fiction rockets and spacecraft probably go in and out of print more rapidly than any other type of publication. There are almost always books, magazines, and even plan packages on sale that deal with the latest motion picture or television shows or with the latest missions of NASA exploration vehicles. Your public library should have a number of the out-of-print books and some magazines. Publications like *National Geographic, Scientific American,* and *Omni* often have the latest photos of current and proposed rockets and space vehicles. Back issues of these magazines can often be extremely helpful if you are constructing a rocket from the near or distant past. There are no regular magazines that deal exclusively with rocket or spacecraft miniatures. Paul Newitt's *Star Fleet Division* booklets describe some very effective techniques for upgrading the fantasy spaceships. The Gamescience "wargame" *Star Fleet Battle Manual* also describes and illustrates variations on similar fantasy vehicles with notes (for the game) on the defense and firepower capabilities of the craft. My own *E.T.V. Model Book*, published by Chilton, has plans for an entire fleet of spacecraft including the "Battle Corvette Nebula" and the adversary ships of the "Ziggurat" fleet. The book also describes painting and weathering techniques as well as most of what you need to know to build flying model rockets. The Centuri *Design Manual* and Estes's *Rocket Plans* series are extremely helpful sources of designs for flying rockets that can also be used as the basis for even more complex static or display model rockets. The "Dou-

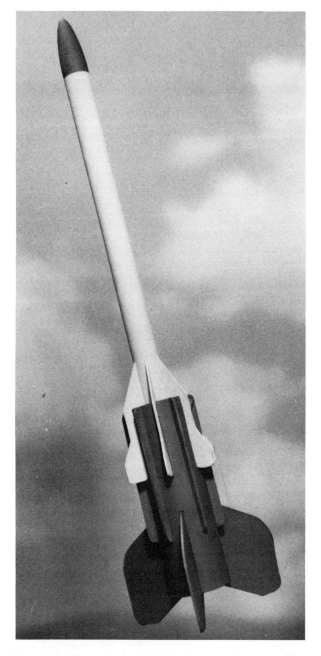

Fig. 2-15 The completed "Double Trouble" with the booster stage *(dark)* and upper stage *(light)* connected for launching.

ble Trouble" rocket described in this chapter is one of the Estes's *Rocket Plans* designs.

Estes and Centuri also have model rocket "clubs" that you can join (for a fee) to be informed of the latest in flying model rocket kits and developments. The monthly magazine of the National Association of Rocketry (P.O. Box 725, New Providence, NJ 07974) is included with membership in this flying model rocket club. If you enjoy rockets that fly, you'll find NAR membership to be most helpful.

Chapter 3

Aircraft

*T*HE hobby of building models of all types has become one of the largest leisure-time pursuits in the world when considered in terms of either hours or dollars spent on the hobby. Aircraft models are, perhaps, the easiest forms of miniatures to use as examples of how a model can help to bring some of your dreams to life. Flight has been one of mankind's dreams since the beginning of recorded time. Most of us have flown as passengers on a commercial aircraft, but the process of piloting your own aircraft has been realized only by a very small percentage of the population. Fewer still can afford to actually own a full-size aircraft, and only a few hundred people in the entire world are wealthy enough to be able to own and fly historic aircraft.

With models, however, you can build and fly virtually any aircraft you can imagine, from a World War I biplane to the most modern military helicopter or jet. If you are on a limited budget, the model can be constructed from scratch over a period of several years. If you have more money than spare time, there are dozens of ready-built flying aircraft models and men who will build kits for a fee. The amount of enjoyment that you can derive from the hobby is certainly not in any direct proportion to the amount of money you spend; some of the most experienced aircraft modelers claim they enjoy building and flying the very inexpensive stick-and-paper "Peanut Scale" flying models as much as the very expensive radio-controlled flying helicopters.

Static vs. Flying Models

You might be surprised to learn that there are far more modelers who build aircraft for display than for flight. The plastic kits that you see in hobby shops and departments under labels such as Revell, Monogram, MRC/Tamiya, AMT, Airfix, Lindberg, Heller, Minicraft, or Testors/Italiaeri are intended only to be assembled and displayed on the shelf. These models are far too heavy and too fragile to be powered with either rubber bands or internal combustion engines. Testors and Cox do make plastic planes that are virtually ready to fly with internal combustion, single-cylinder engines for power. Those models, however, are made from special flexible plastics, and to help reduce the weight of the model, the wings have no lower airfoil surface. Most of the plastic ready-to-fly aircraft models are not precise scale miniatures; the wings must be a bit larger than they should be and the landing gear must be stronger than a true-to-scale model to withstand the rigors of flying and landing.

The popularity of display-only aircraft model kits is an indication that aircraft have an appeal beyond their ability to fly. The airplane, helicopter, or lighter-than-air craft (dirigibles, zeppelins, and the like) have often been the most visible evidence of man's success in his efforts to fly. There are a few aircraft that are quite ugly, but most of them are perfect examples of the form-follows-function

AIRCRAFT

Fig. 3-1 The balsa wood framework for Peck-Polymers' 13-inch wingspan "Peanut scale" flying model of the P-51D Mustang. Courtesy of Peck-Polymers.

theory; aircraft are some of the most beautiful machines created. The aircraft designs of any era, from the kitelike airplanes of the Wright brothers to smooth-skinned jets, are often the era's best surviving examples of man's progress in his battle to master technology. Many full-size aircraft are near-perfect works of the industrial art of their time in the history of flight.

Accurate Replicas

The process of building a miniature of an airplane or an automobile, boat, or locomotive is a two-stage operation: the miniature must be shaped to match the real thing, and the color and texture of the prototype must be captured to the best of the modeler's ability. It is almost as difficult to capture the shape of a real airplane as it is to carve a bust of a human head. Scale-size plans are a help in shaping a model airplane, but it takes more than a perfect replica of the airplane to *look* real. The most skillful industrial sculptors (who carve the master patterns for the molds that make the kits) work from the plans for the actual aircraft whenever they can. Even with those

Fig. 3-2 Folded and curved sheets of precolored paper are cut out to build this Spitfire kit imported by John Hathaway. Photo courtesy of John Hathaway.

plans as a source of the proper shapes, however, these master sculptors must exaggerate some details and shapes and subdue others to create a truly realistic miniature.

Fortunately, most of that work has been done for you when you buy any of the kits that are available today. There is still a great deal of artistic skill involved in assembling even the best of the plastic kits. You, the builder, must see that the parts are fitted together properly with the seams between the plastic covered with putty so that the only visible seams on the model are the door edges, panel lines, and other lines that appear on the full-size aircraft. This process is a bit easier with some kits than with others, but it is always necessary—no kit is "good enough" to be assembled without careful fitting of each part and at least a bit of filler putty to disguise the seams or joints between the separate pieces of plastic.

Every model kit must be painted, even if the kit includes "precolored" parts. The only exceptions to this rule might be some of the paper-covered "stick" models where the paint would add too much weight to the model. Even these flying models can be painted if you carefully "dust-on" almost-dry flat-finish paint with an airbrush. There is not enough space here to discuss all the elements of painting models; a solid chapter or more is needed just to discuss how to paint aircraft, and additional chapters would be needed to describe the different techniques for painting railroad models and still different techniques for painting automobile models. There is also some difference between the methods for properly painting an automobile model that will actually be raced and methods for painting one that is intended only for display. The basic techniques of painting are described briefly in Chapter 1 of this book ("weathering" is described in Chapter 6); you'll need to refer to books that deal with your specific favorite types of models to obtain all the information you should have about painting models.

The net result of any application of paint to any type of model is the same—the miniature is supposed to look like the real thing. The simple application of one or two colors is never enough! Every type of model must receive some additional shading that can range from accenting door edge lines on an automobile with inked-in cuts, to duplicating the weathered appearance of some of the war-damaged and paint-peeled fighter aircraft near the end of the Second World War. If you are going to build a number of plastic display model aircraft or just one or two realistic flying scale models, you'll need to expend some effort learning how to make their surfaces look like the real thing. You must select your colors and apply them with the care of an artist if you want to create a truly realistic miniature.

Display Models

There is a lot more to building a plastic model aircraft kit than just gluing the precolored parts together and slapping on the decals. This particular hobby offers several areas of interest that you can select as your "specialty," or you can combine them all. Many model aircraft hobbyists, for example, spend more time looking through books and magazines for full-size aircraft to duplicate than they do building the duplicates from kits. The history of flight, whether for peacetime transportation or as the development of aerial warfare, is a subject too fascinating to be taught in school. Some modelers begin their research after they have purchased a kit when they want to locate some major or minor variation in the aircraft's color and markings to apply to their model to make it unique. Other modelers try to match their kits to some particular full-size aircraft or squadron or battle that they have discovered while reading about aircraft.

An extremely interesting display can be created by building replicas of, say, a dozen P-51D Mustang fighter aircraft in a dozen of the different paint and marking schemes that aircraft appeared in during 1944. There are several manufacturers of decals for aircraft that provide alternative markings to those in the kits. The decal sheets themselves often provide sources of information on the full-size aircraft as well as its paint schemes as a shortcut in any research program. Some modelers don't bother to research their kit's prototypes at all; they're content to build the models as they appear on the box lid. Others spend much time modifying kits, using two or more kits for conversions to build models that do not appear in kits or to build miniatures (like the Loening OA-1A in this chapter) from scratch.

Scratchbuilt Aircraft

There are several rather diverse types of kits that I can recommend to help you to learn the techniques needed to build a model like the Loening OA-1A or any other aircraft from scratch. First, try to build at least a half-dozen of the plastic kits for display models. One of these should be 1/72 scale, one should be 1/48 scale, and one should be 1/32 scale to give you an idea of the differences in detail and visibility that occur in different size models.

I also suggest that you try to build a conversion of a plastic aircraft kit using parts from two or more kits or one of the vacuum-formed conversion "kits." This might be the time to do some research to discover variations on your favorite aircraft that can be duplicated by modifying a stock kit. The conversion of a single-seat fighter aircraft into the two-seat "trainer" version is one popular example. Some of the other variations of the basically stock aircraft are also fairly simple to carry out. You'll learn the importance of seeing at least two views of the actual aircraft from this conversion exercise and something about making parts fit where they were never intended to fit.

I also suggest that you build at least one of the kits that are produced by forming sheets of white plastic with heat and vacuum (see Chapter 9). The Maserati 5000 GT in Chapter 4 is an example of this technique using clear butyrate plastic rather than the white styrene plastic used in most vacuum-formed aircraft kits. It's just about as easy to carve a master pattern for an aircraft as it is to carve one for an automobile like the Maserati and, if you find you enjoy building vacuum-formed kits, you may want to vacuum form your own using the techniques in Chapter 4.

Building Paper Aircraft

My suggestion for a self-taught "course" in the techniques of scratchbuilding aircraft (or any other models) is to purchase at least one of the paper model kits like the Spitfire (Fig. 3-2). Paper models are often much more difficult to assemble than any plastic kit. The hobby is far more popular in Europe than it is in America, but the majority of the different brands of European paper models are imported by John Hathaway. It is extremely difficult to engineer a model with the curved fuselages and cowls and tires of an aircraft when the basic material is flat paper. The problem is compounded by the fact that all of the paper models are precolored so that no shaping or filling can be used to disguise or smooth any seams. Thus, every joint must be a perfect fit.

Some of the paper models are as simple as the railroad station in Chapter 8; others offer complete interior details and other fine details that are seldom seen in the best injection-molded plastic kits. You may find that the assembly of a paper model is so enjoyable that you'll build most of your miniatures from this type of kit. The experience that you gain in assembling the paper models, however, can be applied directly to the assembly of almost any similar type of model from sheets of white styrene plastic. Some modelers, in fact, use the paper models as patterns to duplicate the miniature in sheet plastic. The process works as well with aircraft as it does with railroad stations. When you duplicate a paper model in plastic, though, you have the distinct advantage of being able to sand and file and fill the seams. Bulbous parts like tires, cowls, and scoops can be carved from solid blocks of plastic with strips of wire or plastic rod used for landing struts and other small-diameter parts. The canopy and windows can, of course, be clear plastic.

The Loening OA-1A

This 1926-era biplane occupies a special place in the history of aviation. Five of these Loening OA-1A aircraft, named the *New York,* the *San Antonio,* the *San Francisco,* the *Detroit,* and the *Saint Louis,* started a four-month, 22,000-mile flight from San Antonio, through most of South America, and back to Washington, D.C. The flight was a diplomatic mission that also served to establish the practical need for passenger flights; Pan American Airways began commercial service over the same general route about two years later (1929). Two of the five aircraft crashed in midair over Argentina, killing the crew of the *Detroit* and destroying both it and the *New York.* Lloyd Jones adapted the No. GPO 908.241 plans, published by the Office of Public Affairs, Smithsonian Institution, Washington, D.C. 20560, in their booklet *The Pan-American Goodwill Flight of 1926–1927,* to 1/32 scale for his scratch-built model. His plans are reproduced here in the more popular 1/72 scale. Lloyd used the seats from two Revell 1/32 scale P-40 kits and a P-40 propeller for his model. The pilot figure is from Monogram's 1/32 scale Gulfhawk kit. The wire wheels are from Peck-Polymers for "Peanut scale" flying models. There are several 1/72 scale kits that can provide similar

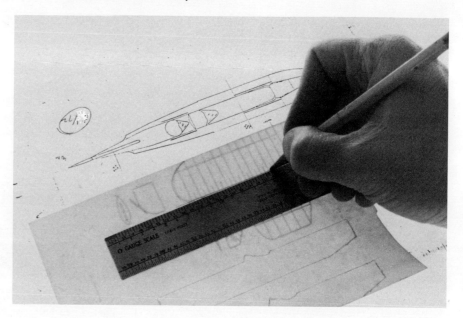

Fig. 3-3 Place the plans on a sheet of clear Plexiglas with a light behind them so you can trace directly onto the plastic.

parts, and the etched wire wheels produced by Precision Miniatures for automobiles would be equally realistic on a 1/72 scale model. If you do want to build the model in 1/32 scale, the plans can be enlarged by 222 percent (1/72 divided by 1/32 = 2.22) by a photostat shop as described in Chapter 1.

Lloyd's 1/32 scale replica of the Loening OA-1A was assembled from .020-inch thick white styrene sheet plastic. Lloyd carved the small pontoons on the wings in two halves and made .010-inch-thick styrene plastic duplicates in a small vacuum-forming machine. The center float was carved from fine-pore polyurethane foam similar to the green-colored plastic used by florists to hold flower arrangements. The float was then covered with a single thin layer of Fiberglas cloth and resin from a boat repair kit. The rounded edges of the resin were filed and sanded to crisp and true contours. The lettering was done with the dry transfers used by a draftsman. Lloyd made patterns for the different colors used on the intricate shields on each side of the plane in pressure-sensitive mailing label paper. He then cut out the patterns and used them as self-sticking stencils to airbrush the colors directly onto the sides of the fuselage. Some of the smaller details were painted by hand with a No. 000 brush.

Using Sheet Plastic

The techniques Lloyd Jones used to build his 1/32 scale Loening OA-1A biplane from sheets of plastic are illustrated on these pages. The most commonly available styrene sheets are the 6½" × 10½" sheets from Evergreen Scale Models. The white styrene plastic from Evergreen is available in .010, .015, .020., .030, .040, .060, and .080 thicknesses and in .005 and .010-inch-thick clear styrene. Evergreen also has a wide selection of scribed styrene sheets and strips ranging from .015" × .040" to .125" × .250" (⅛" × ¼"). You may also find other brands of styrene sheet plastic in hobby shops and artists' supply stores. Many of the "For Sale" signs sold in hardware stores are about .040-inch-thick plastic similar to styrene. Plastruct and Model Parts Inc. offer a range of ABS plastic sheets in gray and clear as well as tubes, angles, I beams, channels, and columns. The ABS plastic is similar to styrene but somewhat stronger, more flexible, and less susceptible to distortion from heat or sunlight. Heat or sunlight will eventually distort and weaken any of the plastics used in models, however, so windowsills are definitely *not* the places to display your models.

The styrene or ABS plastics can be cemented with the same liquid cements or tube-type cements for plastics used to assemble the plastic kits. The tube-type cements should only be used on plastic sheets that are thicker than .030 inch and, then, only in very thin beads so the cement won't cause the plastic to sink and deform as the cement's solvent evaporates.

Use the plans to make bulkheads for the fuselage, but remember to subtract the thickness of the plastic you are using from the out-

AIRCRAFT

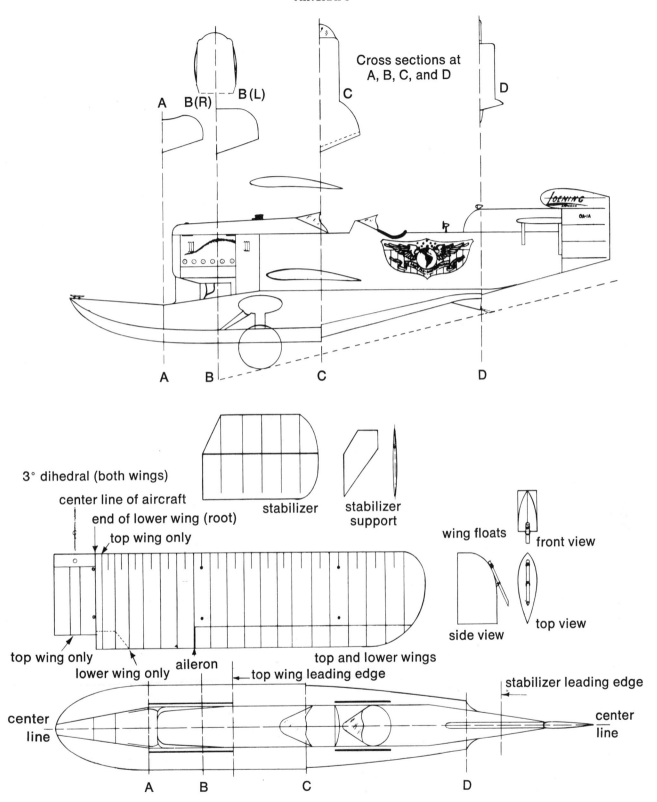

Fig. 3-4 The plans for the Loening OA-1A in the Smithsonian's booklet are typical of drawings of full-size aircraft; they are very small (about 1/170 scale) and do not show some of the details of the wings. Lloyd Jones enlarged the plans and re-drew them to flatten out the wings' dihedral and provide details needed to build the model. His plans are reproduced here in 1/72 scale.

MODEL-BUILDING PROJECTS

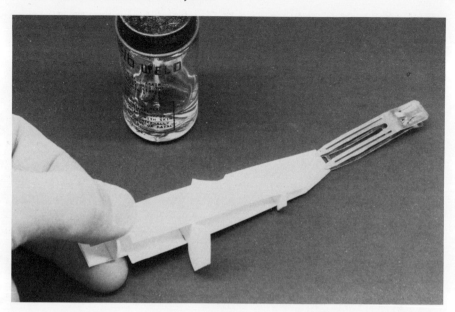

Fig. 3-5 Cut out the sheet plastic sides and bulkheads and cement the parts together with liquid cement for plastics.

side dimensions of the bulkheads. The .020-inch sheets are about right for a 1/72 scale model, but you might want to use .030 or even .040-inch thick sheets for a 1/32 scale model. The bulkheads (and the airfoil or teardrop-shaped spars inside the wings) should be at least .040-inch-thick plastic. The Loening OA-1A had a fabric and sheet metal skin, so many of the panels are flat. The sides of the fuselage, for example, are perfectly vertical with a curve only toward the direction of the tail (see top view).

Begin construction of the fuselage by cutting out a matched pair of side panels, using the plans as patterns. The side panels are then cemented to the bulkheads using liquid cement for plastics as shown in Fig. 3-5. The curved top of the cowl and the top of the fuselage behind the two cockpits can be made from a half-dozen strips of plastic tapered so they fit between the thin tail and wider cockpit. The center float can be made from several .060-inch-wide strips of plastic cemented over bulkheads, or you can carve one the way Lloyd Jones did and glue (epoxy) it to the bottom of the completed fuselage.

Fill any gaps between the outer edges of the styrene plastic fuselage "skin" with automobile body filler "spot" putty, and, when the putty is dry, sand the fuselage smooth. Rivets

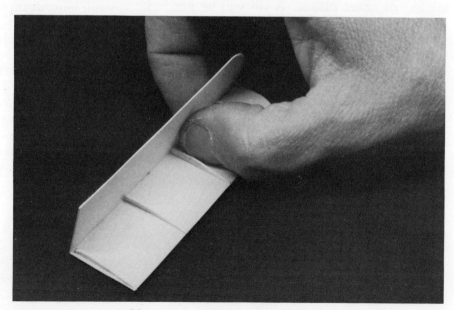

Fig. 3-6 Cement the upper and lower wing surfaces together along the leading edge of the wing with the spars in place.

Fig. 3-7 Fold the upper wing surface over the spars and glue it in place to complete the hollow airfoil for each half of the wing.

and panel lines can be made using the indent-and-paint drop method described in Chapter 9. The gaps between the ailerons and stabilizer and the other hinged-surface seams can simply be scribed into the surface of the model with a sharp hobby knife.

The wings, stabilizer, and rudders of any aircraft are not flat surfaces; they have a thin teardrop or airfoil shape when viewed in cross section. It's possible to sand the smaller stabilizer and rudder to a teardrop shape if you begin with .060-inch-thick plastic parts (for a 1/72 scale model). If the surface is thicker than that, though, it's easier to fabricate a hollow wing, rudder, or stabilizer. The airfoil shape is cut into .040-inch or thicker bulkheads or "ribs" of sheet plastic. A rib will be needed about every ½ inch or so. When building a wing with no taper like that on the Loening OA-1A, cut all the ribs you'll need at the same time and clamp them into a pile or stack so they can all be filed and sanded to shape together. The ribs will have to be shaped individually for tapered wings on more modern aircraft.

Cut the top and bottom panels for the wings, rudder, or stabilizer from .010 or 020-inch-thick plastic but make those parts about .020-inch oversize so that they can be sanded to the correct size when the part is completed. Cement the ribs to the lower panel of the wing with liquid cement for plastics. Use the same cement to attach *only* the forward edge of the upper panel of the wing. The liquid cement must be allowed to dry for 24 hours. Wrap a piece of masking tape over the forward edge of the wing after the cement has dried to temporarily reinforce the cement joint.

Now press the top panel of the wing down so it contacts the tops of all the ribs. Let it spring back up again while you apply two coats of liquid cement for plastics to the tops of each of the ribs, then push the top panel back down and hold it with tape across the trailing edges. Allow those cement joints to dry overnight, and then remove the tape and apply a few drops of cement to the trailing edges of the wing so that capillary action can carry the cement into the seam between the upper and lower wing panels. You may need to clamp the trailing edges with clothespins or small hairpin clips until the cement dries.

Use a bead of automobile body filler to smooth over the leading edge and tips of the wing. When the cement has dried for 72 hours or more, the overall outline and shape of the wing can be filed and sanded to the correct scale size. The assembly of the completed wings, fuselage, stabilizer, and rudder is just like the assembly of any plastic kit except that you may need to use a little more filler putty if you didn't fit each part together perfectly.

Flying Aircraft

It is possible to use a technique similar to that just described for the display model Loening OA-1A to build a flying model aircraft. The flying model, however, would be constructed from sheets of light-weight balsa wood for all the bulkheads and spars with a choice of either

Fig. 3-8 Lloyd Jones' replica of the Loening OA-1A is a 1/32 scale model that has won several International Plastic Modelers Society contests. The parts he used are also available from 1/72 scale kits. You might find it a bit easier to duplicate his efforts in 1/72 scale because slight mistakes in alignment or detail won't be quite as noticeable. Lloyd used hand-cut paper masks to spray paint the markings.

thin plywood, tissue paper, or special plastic coverings for the outer surfaces. Unfortunately, if you build a flying model from plans of a full-size aircraft, there is a better than even chance that your model won't fly properly. It takes years of experience flying and building model aircraft to determine just how much wing area, how much control surface area, and how large an engine and propeller must be used to achieve a balanced design that works on a model. If you want to build a model from scratch, then, use some of the plans that have been drawn by model aircraft engineers to build scale model replicas of your favorite aircraft. The magazines that deal with flying aircraft models usually publish plans in every issue. There are also ads in these magazines for currently available sources of flying aircraft plans, and you'll find others in the back of this book. Some of the plans are merely exterior views of models that have been built and flown successfully, but most of the plans include full-size drawings of the bulkheads, spars, and other interior structural parts of the model along with a bill of materials. Do not attempt to reduce or enlarge the plans for a flying model as you might for a display model; the flying model will fly as it was designed *only* in the same scale as the plans. Larger or smaller models will require changes in the engine size and in the sizes of some of the control surfaces to fly properly.

Flying Model

Some special skills are needed to build an airplane model that flies. The wooden inner structure must be assembled so that it remains in perfect alignment even after the glue dries. Only the minimum amount of glue needed for a permanent bond should be used to keep weight to a minimum, and, finally, the outer skin of the aircraft must be perfectly smooth. The best way to learn these techniques and the others needed to build aircraft models that fly is to assemble at least three or four flying aircraft kits. There are literally thousands of flying aircraft kits that range from almost-ready-to-fly (ARF) models, made mostly of foam plastic, to boxes of wood, paper and plans to build the "stick model" kits. If you are more interested in flying than in building, I recommend one of the almost-ready-to-fly models.

If you want to learn to build, then start with one of the simple "Peanut Scale" models from which you can build an aircraft with a simple rectangular or triangular cross-section fuselage. The kits for larger scale models and any type of kit for a model with a round fuselage can be quite difficult for a beginner to build. Even the relatively simple Peck-Polymers P-51D Mustang shown earlier in this chapter should be reserved for your second or third attempt at building a flying model from a kit. Most of the kits have die-cut ribs and bulkheads, to make assembly a bit easier. Don't let such precut pieces fool you; some of those models with 36-inch and longer wingspans can take hundreds of hours to finish.

You will need to decide how you want to power your flying model and how you want to control that flight before you even purchase a kit. There are basically four choices of power: rubber bands, internal combustion–type engines, jets (or ducted fans), and electric motors. Internal combustion engines are simple sources of power for the almost-ready-to-fly aircraft, but they can add some incredible complications if you want to fit them in a "stick" model (regardless of what the kit box might say). The rubber-band powered aircraft are relatively easy to build in the peanut scale, but they can be incredibly complex models in larger sizes. Jet or ducted fan engines are best reserved for modelers with several years of flying and building experience because the models' flights are somewhat difficult to control. The electric motors provide silent flight that can be a real advantage if you want to at least test fly the model in town. The electric motors are powered by rechargeable batteries carried inside the aircraft. It takes a real expert to design a model that will fly well with that extra load, so you'd better build kits that are designed for electric motor power—don't attempt to replace an internal combustion engine or a rubberband-powered engine with an electric motor until you've had experience building several electric power aircraft models.

You may be wise to consider a sailplane for one of your early models. The sailplanes are carried aloft by 100-foot-long rubber bands and cables or by low-power (for the size of the model) internal combustion engines or electric motors. The engines or motors are designed to carry the model only to the peak of its flight when they shut off so that the model can glide or "sail" through most of the duration of its flight. Sailplanes are specialized designs that are either replicas of full-size sailplanes or special configurations made only as models—most have relatively long and thin wings and fuselages compared with conventional power planes. Some other models' flight can be controlled either by two long cables or lines ("U-control") or by a radio transmitter held in your hand with the receiver in the aircraft ("radio control"), or the model's control surfaces can be adjusted to fly in wide circles with no outside control ("free flight"). Again, the model must be one that is designed for the type of control system (as well as the type of power) that you desire; only an expert should attempt to fit radio control, for example, into a model designed for free flight.

Flying Stick Models

The classic flying model airplane is one that is built from sticks and sheets of balsa and covered with tissue paper. These were the airplanes that established the sport and hobby of model airplanes back in the thirties. There's a strong revival of "stick" model building today, thanks to what are called Peanut scale models. Most of these aircraft have a wingspan of about 13 inches with the exact scale of any prototype-based models adjusted to match the wingspan.

Some of the Peanut scale models are freelance designs which are not based on any actual full-size aircraft; these models, like the miniature in Fig. 3-9, fall into the "Peanut sport" category. Lou Roberts designed this Peanut sport model to be a perfect first-time model. The fuselage, wings, rudder, and stablizer are all easy to cover and they have built-in strength, thanks to special gussets and the fuselage's triangular shape. Best of all, this model should fly as well as anything in its class if you match it exactly to the plans and pay attention to every notation on the plans.

When you're building a model from scratch, you cannot assume anything; every step and every subtle gusset or angle must be matched precisely to the plans. You can build this plane using the parts from most of the Peck-Polymers or Micro-X Peanut class model kits, or you can simply purchase the materials. Only a few hobby shops stock true Japanese tissue, the lightweight balsa that works best with

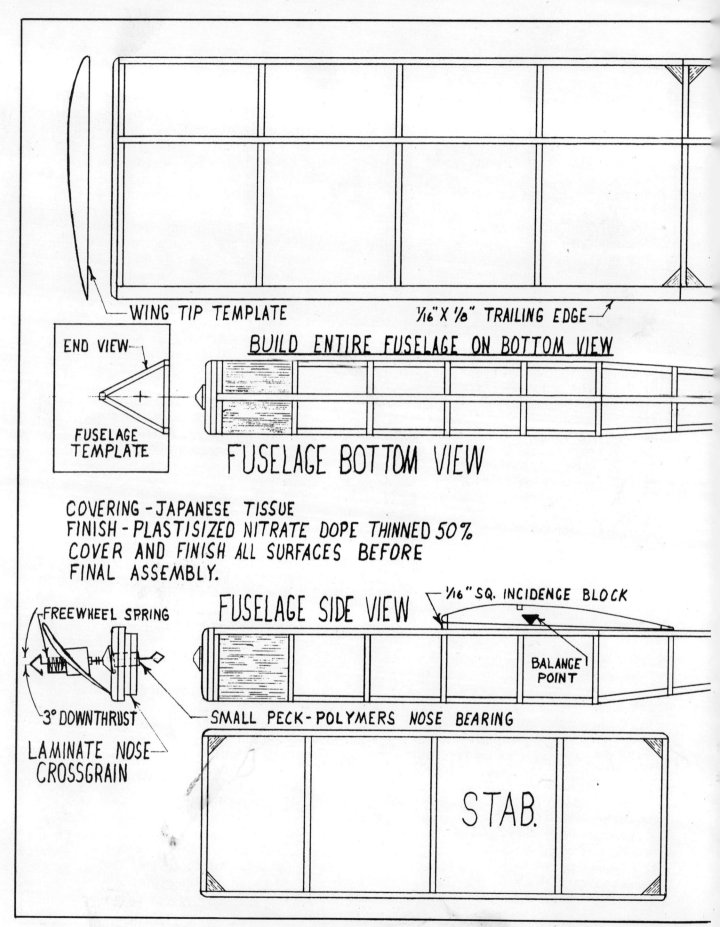

Fig. 3-9 Full-size plans for the "Peanut sport" flying model airplane. Courtesy of Lou Roberts.

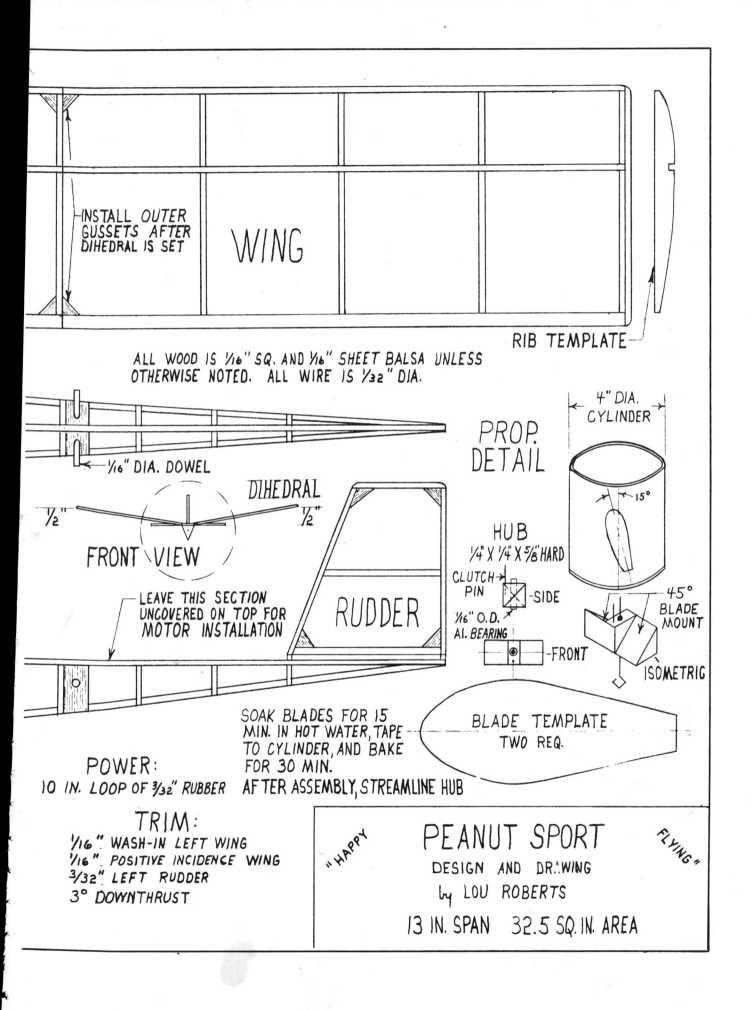

such tiny aircraft, the correct type of 3/32-inch rubber band material, or the Peck-Polymers nose bearing and propellor thrust washers. You may simply want to order all of the materials for this first model by mail or adapt a kit's parts. Make a copy of the plans full-size so you can tape it over a sheet of Celotex or similar soft wallboard. Cover the plan with waxed paper and you're ready to scratchbuild your first flying airplane model.

Getting Up and Away

Use a sharp hobby knife to cut the balsa parts to match the plan. Carefully cut the two propeller blades from 1/16-inch balsa sheet, soak them in hot water for 15 minutes, and tape them to a 4-inch-diameter can (cylinder) as shown on the drawings (Fig. 3-9). The propeller hub must be carved from a 5/8-inch long piece of 1/4-inch square hard balsa. You can drill the 1/16-inch hole for the bearing and a 1/32-inch hole for the clutch pin (used with handcrank winding machines) with drill bits held in your fingertips. Aluminum tubing of 1/16 inch diameter is available from most hobby shops; cut it by slicing around the diameter with an old hobby knife blade, and then snap the tubing off by holding the cut line over the edge of a table. Bend the 1/32-inch steel piano (music) wire hook for the propeller and rubber band using pliers. Do not try to cut the piano wire; bend it back and forth on the "cut" mark with two pair of pliers until it breaks. I used the Goldberg brand Super Jet thickened cyanoacrylate cement to assemble all the wood parts, but plastic resin-based cements like Franklin Chemical's Tite-Bond work just as well although they dry slower. You can also use any of the thinner cyanoacrylate cements (like Hot Stuff) if you sprinkle a touch of baking soda on the joints to serve as a filler material for any slightly misfitting parts. The cyanoacrylate cement will bond the baking soda into almost rock-hard filler material. The paper cover should be attached (after the model is completed) by spreading white glue diluted with an equal part of water over the framework of the model.

Cut the ten wing ribs (to match the rib template) from the 1/16-inch balsa wood sheet. Cut two triangular pieces of the same sheet to match the outside shape of the fuselage (the fuselage template) and two triangular pieces to match the smaller inside of the fuselage's triangular shape. Glue the four 1/16-inch balsa wood triangles together in a stack, rotating the grain of the wood 90 degrees on each piece for strength. When the glue is dry, drill a 1/8-inch hole to match the small Peck-Polymers' nose bearing. Locate the hole at the "X" indicated on the end view of the fuselage template and be sure to angle the 1/8-inch hole downward 3 degrees, as shown in the side view of the aircraft's nose.

The propeller blades, hub, and 1/16-inch aluminum tube bearing can now be cemented together. Do *not* cement the Peck-Polymers' nose bearing into the nose of the fuselage; you

Fig. 3-10 Sprinkle some baking soda at each joint if you are using cyanoacrylate cement (like Hot Stuff) to assemble the model.

may want to replace the bearing if it is worn. The freewheel spring takes up the tension in the rubber band when the rubber band is slack. This keeps the propeller from sagging or wobbling and upsetting the stability of the glide portion of the aircraft's flight. Cut the spring from the softest coil spring you can find inside a ballpoint pen. The spring's tension should be less than that of the half-wound rubber band. Assemble the propeller with the two Peck-Polymers' brass thrust washers between the protruding bit of 1/16-inch aluminum tubing (that sticks out from the rear of the propeller hub) and the face of the Peck-Polymers' nose bearing. Bend the 1/32-inch diameter length of steel piano wire to the exact length and shape shown in the drawing to retain the entire assembly in the nose. The nose can be covered with tissue paper to match the aircraft but *never* glue the nose in place—you may want to adjust its angle to vary the thrust angle of the propeller to help "trim" the aircraft's flight pattern.

The triangular fuselage should be assembled on the fuselage bottom view template of the plans. Cut the 1/16-inch square balsa ribs to length right on the plan and pin them in place over the waxed paper (see Fig. 3-10). Add a drop of glue to each of the joints between the pieces of wood. When the top of the fuselage is completed, cut another 11 *pair* of cross ribs (bulkheads) from the 1/16-inch square balsa. Use these ribs or bulkheads to brace the single two-piece rib that forms the bottom edges of the triangular fuselage.

The rudder is shown attached to the fuselage in the plan; however, those 1/16-inch strips between it and the fuselage are actually an end view of the stabilizer (marked "stab." on the plan). Do *not* glue the rudder to the fuselage. Use the plan to assemble the rudder just as you did for the top of the fuselage, but use only a single strip of 1/16-inch square balsa along the edge of the rudder that joins the stabilizer. When the fuselage, wings, rudder, and stabilizer are all complete *and* covered with tissue paper, then the stabilizer will be glued to the top of the fuselage with the rudder glued to the top of the stabilizer. Use the plans as a jig to complete the 1/16-inch square balsa framework for the stabilizer and to assemble the wing with the ten ribs, two wing tips, and the 1/16" × 1/8" trailing edges—the leading edge and the top spars of the wing are 1/16-inch square balsa. The wing is to be made in three pieces: the center section (which will be immediately above the fuselage) and the right and left wing halves. Glue the three pieces together with the wings angled upward to give 1/2 inch of dihedral, as shown in the front view, but do *not* glue the wing to the fuselage until both are covered with tissue paper.

Tissue-Covering Techniques

The larger scale models intended for outdoor flight are usually covered with a paper-thin sheet plastic material like Top Flite Models' Super Monocote, Pactra's Solar Film, or Polks' Wing Skin. These plastics are designed to be placed over the model's balsa wood structure and heated with a special version of a home laundry iron. The iron softens a built-in cement to stick the plastic to the wood and, at the same time, stretches it to a drumlike tightness over the model's surface. These plastics are far too heavy for any model with a wingspan less than about 24 inches; the smaller models should be covered with the traditional (since the twenties) Japanese tissue paper.

To cover a model with Japanese tissue paper, cut the tissue into blocks that are at least 1 inch larger than each of the flat panels of the model. Each side of the triangular fuselage should be covered with a separate piece of tissue. Work only with one panel at a time, spreading a 50/50 mixture of either white glue and water or plastic resin-based glue (like Titebond) and water. Stretch the tissue slightly between your hands and press it down over the glue-coated surfaces. Allow the glue to dry before adding another panel of tissue. You can cover the tops of the wings, stablizer, rudder, and one side of the fuselage in one evening, another side in a second evening, and the third side in a third evening. Do not cut the paper to match the size of the panels; use a piece of fine-grade sandpaper glued to a scrap of wood (see Fig. 3-11) to sand through the tissue to the wood beneath it (actually, just to the coating of glue).

When the model is covered, it should be spray painted with a 50/50 mixture of Sig's Lite-Coat clear paint and Sig thinner. Use one of the hand-pump bug spray guns if you do not have an airbrush. The markings (such as the peanut on this model) can be applied with a felt-tip marking pen *after* the clear paint has dried. If you want to add colors to the model to duplicate a World War II camouflage pat-

Fig. 3-11 When the individual sheets of tissue paper are cemented to the model, trim the paper to size by sanding away the excess paper.

tern, for instance, the flat-finish paints can be used with an airbrush. Spray the paints right from the bottle using a wide-open paint flow setting on the adjusting knob or lever and about 25 psi of air pressure. Keep the airbrush a foot or more from the model so that the paint will be almost dry when it strikes the model. In essence, you are almost covering the model with a slightly damp "dust" of color. This is a rather tricky procedure that should only be attempted after you have some experience using an airbrush to paint display models.

There is no way to apply paint with a brush without adding too much weight to the model or softening the tissue so that it wrinkles and sags—you must use an airbrush or be satisfied with the original color of the tissue. The model in the photographs was painted only with the

Fig. 3-12 Cover all the parts before assembling them to complete Lou Roberts' scratch-built "Peanut sport" flying model.

50/50 mix of clear paint and thinner to seal the tissue paper so it would not be affected by moisture.

Trim Adjustments for Flight

When the model is completely covered with the tissue, the wing, fuselage, stabilizer, and rudder can be cemented together exactly as shown in the plans. Note that the rudder should be glued in place with the leading edge angled 3/32 inch to the right (left rudder) to force the aircraft to circle in a counterclockwise direction. The tip of the wing is supported 1/16 inch above the fuselage to give 1/16 inch of "positive incidence," and the left tip of the wing must be warped 1/16 inch in a clockwise direction (in relation to the wing section at the fuselage).

These "trim" adjustments should result in a perfect gliding flight indoors (in a gymnasium). Your model may require slight variations in any of the trim adjustments. If your model does not balance with the center of gravity at the "balance point" indicated on the plan, tiny slivers of lead shaved from a fishing weight can be glued inside the front or rear of the fuselage.

Chapter 4
Automobiles

MINIATURE automobiles are, perhaps, the most exciting types of models. Model cars, for both display and actual racing action in miniature, have been the best-selling models for more than two decades. This segment of the hobby encompasses as wide a range of miniatures as the aircraft, rocket, armor, or railroad hobbies. The hobby offers a variety of kits that are intended solely for display as well as an incredible array of miniatures for racing on a tabletop ("slot" cars), and the extremely popular range of radio-controlled racing cars.

The sizes of the more popular model cars range all the way from postage stamp–size 1/160 scale models to 1/8 scale giants the size of a telephone book. Hundreds of new kits are introduced each year that add the very latest full-size cars to the shelves as well as provide new replicas of some popular older automobiles.

Trucks and motorcycles are included in this category as well. The range of motorcycle miniatures is nowhere near as extensive as the range of automobiles. You may have to search some of the foreign magazines for ads of motorcycle kits not imported into America or purchase automobile or truck kits that include a motorcycle as an accessory. The number of truck kits, however, has increased considerably over the past few years with a growing number of large 1/24 and 1/32 scale highway tractors and trailers and similar miniatures in 1/87 scale (for use on HO scale model railroads).

It's relatively easy to create a slot racing or radio-control replica of most of the display model kits by making a vacuum-formed clear plastic replica of the body to mount on a standard chassis. Similarly, many clear plastic bodies intended for slot cars or radio-controlled cars can be superdetailed and mounted on chassis from similar-sized static models. This interchangeability increases the range of available automobile models so that it encompasses a miniature, in some scale, of just about every exciting car or truck that has every existed.

Plastic Model Kits

The injection-molded styrene plastic model automobiles are the most readily available kits on the market. The same construction procedures outlined in Chapter 2 for rocket models and painting techniques outlined in Chapter 1 apply, of course, to the plastic automobile kits. The range of metal molds to produce plastic automobile models is several times larger than that which exists to produce plastic aircraft kits. Several model kit manufacturers have introduced replicas of each year's full-size automobiles to coincide with the introduction of the real vehicles in dealers' showrooms. These "annual models" have been a part of such brands as AMT, JoHan, and MPC/Fundimensions for over 20 years.

It would be impossible for so many different kits to remain in production, so each year several of the previous year's model kits are dropped from the line to leave room for the newer models. If the full-size automobile proves to be a "classic," like the 1965 Corvette, then the kit may be reintroduced into the line. The vast majority of the "annual" models, however, become true collectors' items after they have been on sale for only a year or two. Most automobile model enthusiasts follow the same buying patterns as aircraft model enthusiasts; they purchase as many kits as they can afford the moment the kit is on the market.

Metal Auto Miniatures

There has been an incredible increase in the range of automobile, truck and bus models, as both kits and ready-made miniatures, in 1/43 scale. These cars were first made popular with the zamac, mazac, or zinc metal automobiles from firms like Corgi, Solido, Rio, and Matchbox (Models of Yesteryear series). There are now several dozen firms around the world producing these ready-made models for both the toy and the collector's markets. There are also several dozen manufacturers who produce 1/43 scale metal kits with the parts cast in what are called "white metal" or "soft metal" alloys of tin and other trace metals.

The list of 1/43 scale kits grows weekly, and many of the models are rather limited in their appeal so they are only produced once. Even when these limited-run kits are subtracted from the list of available kits, however, there are still nearly a thousand 1/43 scale model car kits on the market in any given month. Probably several times that amount are now out-of-production collectors' items. A few of the new kits are listed in the car magazine *Road & Track* each month, and others appear in the British monthly *Scale Models*. The only way to keep track of the majority of the new 1/43 scale models, however, is to subscribe to the newsletters issued (for a fee) by the various manufacturers and importers of 1/43 scale kits. There are also a number of ready-built and kit models in metal from firms like Monogram, Ertl, Gabriel (Hubley), Burago (Italy), and Wills Finecast (Britain) in 1/24 scale as well as a few metal models in 1/16 scale and some other in-between sizes.

Building a Metal Model

The tin-based alloys used on most of the 1/43 scale kits are soft enough to be cut and filed like plastic. The metals used in these kits have an advantage over plastic in that they can be bent slightly to align any parts that are warped enough not to fit together properly. Do not, however, attempt to bend the rock-hard zinc alloys used in the Monogram, Gabriel (Hubley), or Ertl 1/24 scale kits because they will break. Most of the kits made of zinc metals fit together a bit easier than the "soft metal" kits after the excess mold flash is filed from the edges of the parts. The zinc metal kits are generally designed to be assembled with a few self-tapping screws with smaller parts that snap into place. It's still wise to use a drop or two of one of the cyanoacrylate instant cements to hold these smaller parts.

It's not quite as easy to make a general statement about the fit or the 1/43 scale kits' parts;

Fig. 4-1 The Wills Finecast 1/24 scale M.G. K3 has the firm's unique real-wire wheels. Ron Klein built this kit.

some of the kits require a considerable amount of filing, bending, and even filling with putty to get the parts to fit together snugly. In general, plan on spending much more time fitting the parts in a metal kit than you would with a plastic model of similar size. Once the parts fit properly you can use either 5-minute epoxy to hold them together or one of the "thick" cyanoacrylate cements such as Goldberg's Super Jet. The kits seldom fit well enough to allow the use of conventional cyanoacrylate cements, and no other glue has the strength to hold the relatively heavy metal parts together.

Car Conversions

There are four major areas of detail in any model car or truck: the visible exterior or body, the engine area, the chassis, and the interior. Very few of the ready-built or kit automobile and truck miniatures have all of these areas detailed properly. Some of the snap-together kits and ready-made metal models, for instance, have very well-shaped bodies but either crudely molded or nonexistent engine or chassis details. When you have purchased a dozen or so kits, from different manufacturers, you'll discover that some of them do a much better job in one area of detail than in another. Experienced model builders will often purchase an entire kit just to obtain especially well-detailed wheels and tires.

Many of the plastic model kits include extra details that may be useful on other models. The automobile modeler soon collects a rather large "scrap box" of leftover parts that frequently prove to be very useful in superdetailing automobile models since there are almost no sources of "accessory" details like wheels, tires, engines or chassis. Precision Miniatures is one of the few firms that offer "accessory" wheels and tires. Their range of 1/43 scale wire wheels are exquisite models in their own right with the double row of "wire" spokes etched (see Chapter 11) from brass and plated to look like chrome. A single detail, such as precise-scale wire wheels, can transform a "toy" model like the Matchbox M.G. TC (Figs. 4-3 and 4-4) into a much more realistic miniature. The "real wire" spokes in the 1/24 accessory wire wheels from Wills Finecast (Fig. 4-1) can add a similar touch of realism to

Fig. 4-2 Leonard Frere used two Athearn 1/87 scale plastic Freightliner truck kits to make this flatbed conversion. Monogram makes a 1/32 scale kit that could be used for such a conversion and AMT has a 1/25 scale model.

Fig. 4-3 The parts for the Precision Miniatures "20-inch" etched-brass wire wheels that will replace the crude cast wheels on the Matchbox M.G.

1/24 scale classic automobiles and smaller scale motorcycles.

The range of automobile models is so extensive that very few full-size automobiles have been overlooked by the kit manufacturers.

Therefore, if you are going to create a special miniature of your own, make a thorough search of both the current production and the "collectors'" models to be certain that your dream car doesn't already exist as a kit or

Fig. 4-4 The delicate spokes in the Precision Miniatures' wire wheels can make even a "toy," like this M.G. TC, look like a scale model.

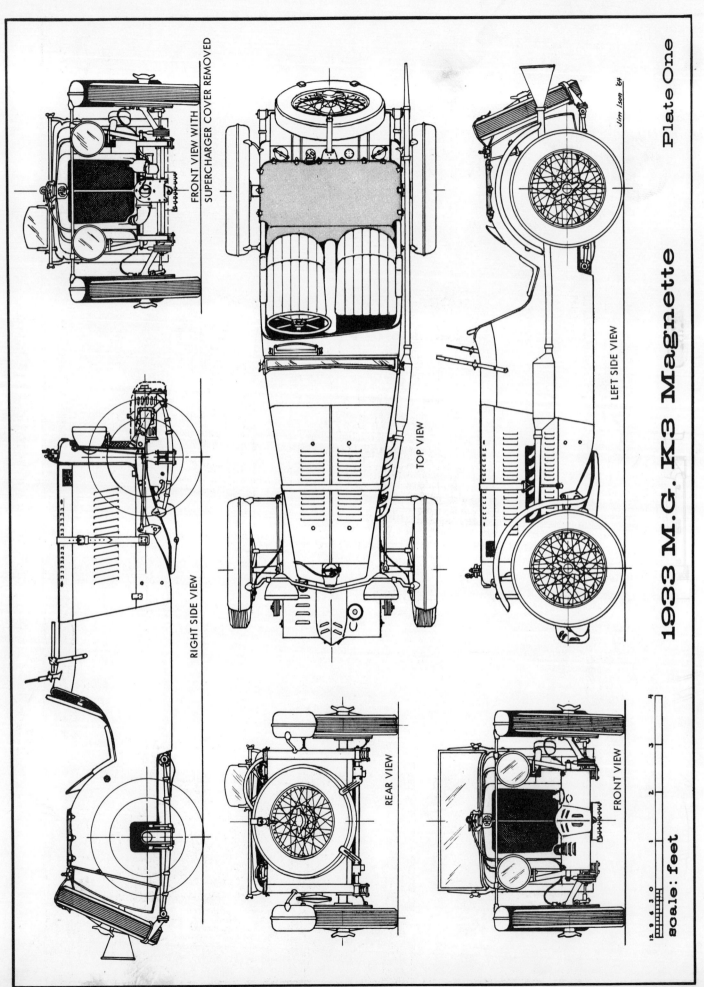

Fig. 4-5 The external views of the 1933 M.G. K3 in 1/24 scale. Drawings by Jim Ison. Copyright © 1964 Autoplans.

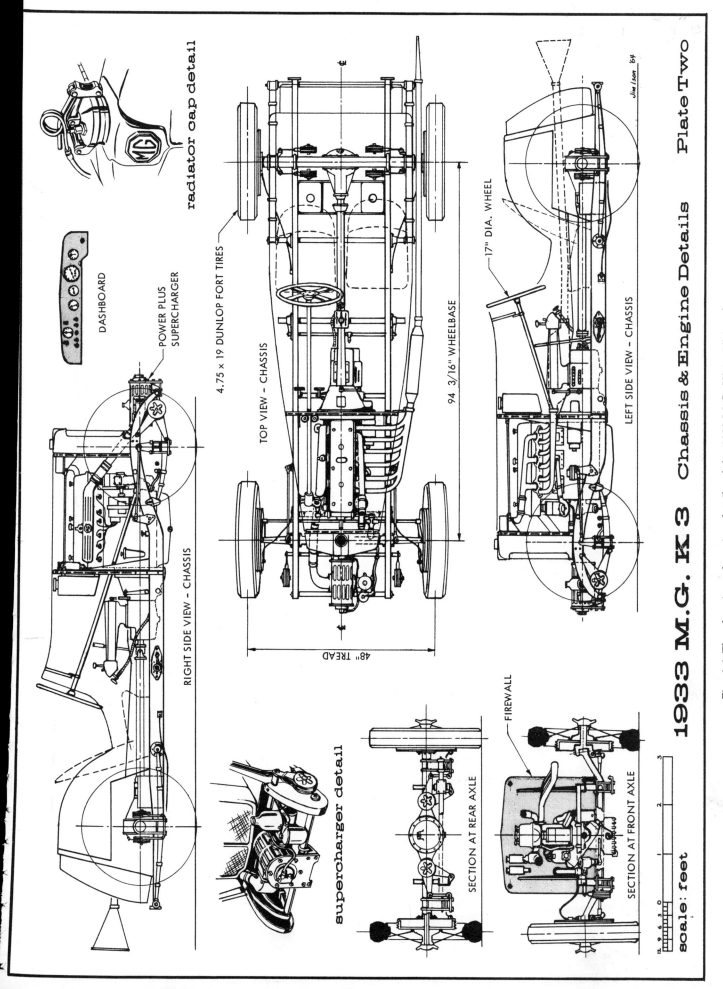

Fig. 4-6 The chassis and engine details of the 1933 M.G. K3 in 1/24 scale. Drawings by Jim Ison. Copyright © 1964, Autoplans.

ready-built model. Hundreds of pre-1970 racing cars were once offered as slot car kits during the 1960–1970 period. The bodies from these models can easily be adapted to and from the static model plastic kit's chassis for 1/32 or 1/24 scale (the two popular sizes for slot cars) to create a unique display model.

Conversely, you may want to use one of the plastic or metal model car bodies as a mold to vacuum form a clear plastic body for a 1/32 or 1/24 scale slot car or for a 1/12 or 1/8 scale radio-controlled racer. The injection-molded plastic bodies can be adapted directly to slot car chassis as shown in Chapter 8. Many of the bodies from the "snap-together" kits are particularly well-suited for use on slot cars because the bodies are mostly in one piece. A well-detailed "snap-together" body can also be used with a detailed chassis from another kit to create a more realistic display model.

The Automobile Museum

I've included James Ison's drawings for the 1933 M.G. K3 Magnette (Figs. 4-5 and 4-6) in 1/24 scale even though there is an excellent "soft metal" kit available from Wills Finecast (Fig. 4-1). These drawings show a number of details (such as fuel lines) that can be added to the details in the kit to create a superdetailed version of the K3. The drawings will also be useful to those who might want to scratch-build a K3 in some other scale or convert some other kit. There are no 1/43 or 1/32 scale kits for the 1933 K3, although there are kits for the 1934 long-tailed version.

The AMT/Matchbox 1/32 scale M.G. TC and the Monogram 1/24 scale and Entex 1/16 scale TC kits could furnish the chassis, grill, hood, and wheels to give you a good start on a kit conversion K3. James Ison's drawings include a number of cars of this vintage as well as cars from the fifties (like an Allard J2) and the sixties (like a Chaparral). The Model and Allied Publications plans include cars up to the most modern Grand Prix racers. Most experienced car builders, you see, settle on one particular scale for most of their projects so that the completed "collection of models" is really a miniature automobile museum. This is why the relatively smaller scales like 1/43, 1/32, and 1/24 are the most popular for automobile miniatures. A dozen or so miniatures can fit on a single shelf of a bookcase if they are all built in the popular scale sizes. If you have a particular favorite automobile, you might want to build a replica in 1/16, 1/12, or even 1/8 scale, but few of us have the space for more than two or three such giants.

Hand Carved Car Bodies

The exterior shape or body is the most appealing aspect of any automobile, whether full-size or miniature. The most difficult part of building a miniature automobile from scratch is achieving the correct sculpturing of the body shape. Plans alone are not enough; you must develop the skilled eye of a sculptor to capture the subtle curves and bends that give each full-size automobile its unique character. Wood is an extremely popular material for sculptors but, frankly, it's one of the most difficult materials to work with when you are carving something with precise details such as door lines, wheel wells, and grill openings. Such details as door handles, grills, and even windows and frames can be added to the model rather than carved into the shape, but that doesn't lessen the need for crisp detail in the body itself. The Cub Scouts and other similar youth groups often have "pinewood derby" races for hand-carved wooden automobiles that are merely allowed to coast down a ramp or hill. The plans for the 1964 Ferrari Formula I car (Fig. 4-7) can be used, with the wood-carving techniques in Chapter 10, to produce a variety of sleek, cigar-shaped "pinewood derby" cars like those in Fig. 4-8. Most of the midsixties era open-wheeled Grand Prix cars carried very similar shapes, so the Ferrari can be subtly modified to create a Lotus 25, a Brabham, or a BRM of the 1964–1965 racing seasons.

The Maserati 5000 GT Model

The Italian-made Maseratis were a major force in world racing during the fifties and sixties. The 1964 Maserati 5000 GT was one of the factory's last efforts to win the prestigious LeMans 24-hour endurance race in France. The chassis for this car was a revised 1962 framework fitted with the sleek flat-tailed body that was the ultimate in aerodynamics before wings and ground force theories altered the shape of such race cars into the boxlike contours of today. Only one car was ever constructed, and it failed to finish any of the three races in which it was entered. The car was later converted to use on the street with a somewhat longer wheelbase and an even more

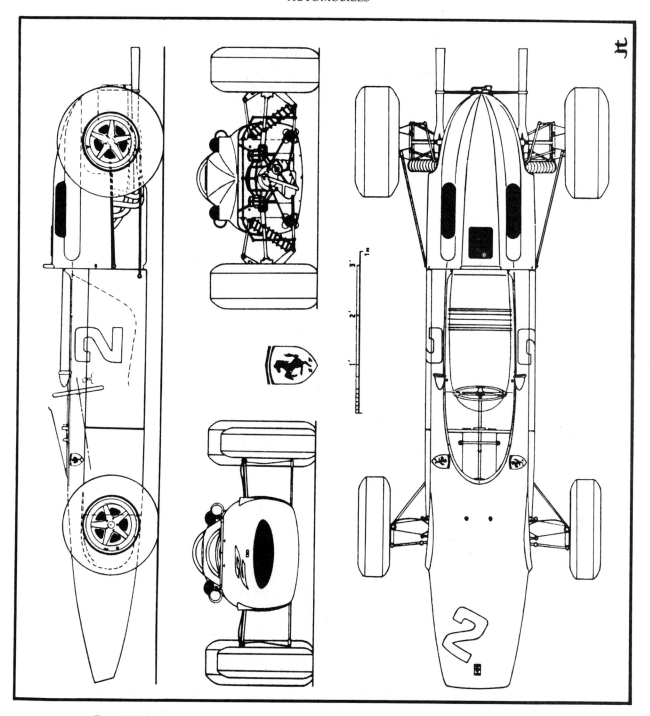

Fig. 4-7 The 1964 Ferrari Formula I Grand Prix car in 1/24 scale. The profile views of the body can be used (see Figs. 4-8 and 10-1 through 10-7) to carve a wooden replica of this automobile with either wooden axles or detail parts from a 1/24 scale plastic racing car kit. Drawings by Jonathan Thompson.

streamlined (though most think uglier) body. A company called Shark Racing Bodies once produced a 1/24 scale clear plastic slot car body of the 5000 GT and Hawk used to sell the 1/30 scale (they call them 1/32 scale) injection-molded plastic display model kits.

If you want a model of this car today, you're going to have to carve it yourself—the Shark bodies and Hawk are virtually impossible to find because they have not been produced for 15 years. Ron Klein carved the 5000 GT shown in Figs. 4-11 through 4-13. He was generally

Fig. 4-8 Pinewood derby cars; a Lotus (No. 30), BRM (No. 8), and a Ferrari all shaped from variations of the Ferrari's plans.

acknowledge to be one of the best automobile sculptors in the world until a stroke crippled him in 1979. There is no better teacher, and most of what you'll find on these pages are the techniques Klein used in sculpting his masterpieces.

Clear Plastic Bodies

It is possible to use the hand-carved "master" for the body of a display model car if you hollow out the interior as shown in Chapter 10. That's the hard way and you take the risk of damaging the master. If you have the skill to carve your own model car bodies, you should have no trouble learning the techniques of making a vacuum-formed clear plastic replica of your original pattern. Alternatively, that pattern can be used to make a mold of RTV rubber for making replicas of Fiberglas cloth and resin as described in Chapter 12. Some professional model car builders, like Gerald Wingrove and Michael Conti, carve a wood body and use that pattern as a "buck" or backup to hammer out body panels from sheet brass or sheet aluminum. If you want to try this, I suggest you enroll in at least one semester of automobile body repair at your high school or trade school—the techniques of heating, annealing, shaping with hammer and block, and filling with putty are almost identical whether you're working with a full-size automobile body or a scale-model metal replica.

A vacuum-formed body, made from clear butyrate plastic, is a much more logical choice for a 1/24 scale or smaller model. You can make your own body using Idea Development's Formicator vacuum-forming machine, or you may be able to persuade a local vacuum-packaging (sometimes called "skin wrap") firm to "pull" a few bodies for you in their giant machines. The clear plastic body provides the most dif-

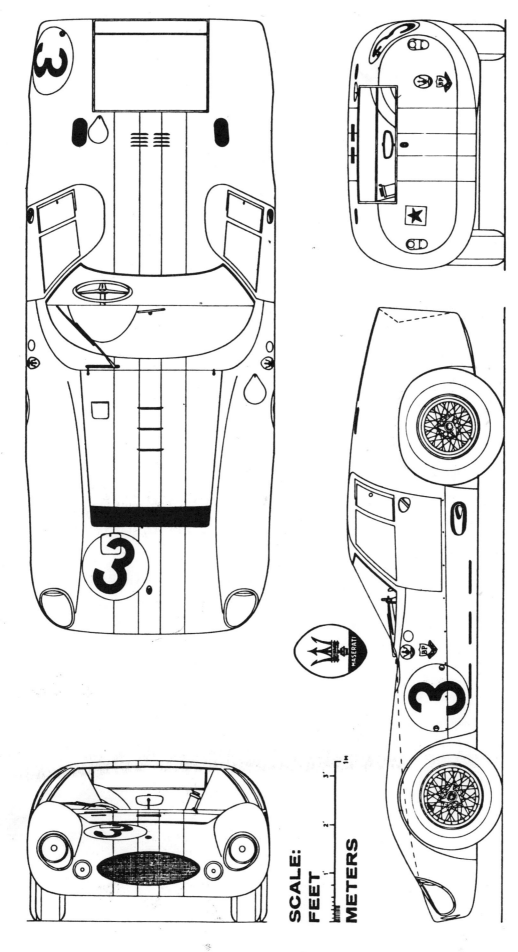

Fig. 4-9 The Maserati 5000 GT prototype as raced at Reims, France, in 1964. The top, side, and end profiles of the body are all that are needed to start carving a plaster (or wood) replica of this classic racing car. Drawings by Jonathan Thompson.

MODEL–BUILDING PROJECTS

injection-molded plastic display model car kits.

Vacuum-Formed Bodies

The master pattern for a vacuum-formed body must be protected from the heat used to soften the plastic sheets before they are formed over the body pattern. The only materials that are suitable as-is are metal and some of the metal-filled epoxy resins; both require the use of a milling machine and grinder to carve and shape the body. Common plaster of paris is probably the best material to use for the basic body. Use an old cardboard box to make a mold for a block of cast plaster that is at least ½ inch larger than your proposed model in every dimension. When you pour the plaster into the cardboard, bang the cardboard on the table (like a drummer) to shake any air bubbles in the plaster to the surface. There will be less chance of air bubbles forming if you mix the plaster thoroughly, adding plaster to the water (never water to the plaster) until the mixture is about the consistency of thick cream. Trace a profile view of the body onto the block of hardened plaster and use a jig saw to cut about 1/16 inch outside that outline. Repeat the process to cut the outline of the top view of the body into the plaster block. You can even use the jig saw to round off the edges of the block if you are careful to stay at least 1/16 inch outside the final profile or shape of the body. The final shape can then be carved using hook-shaped blades for convex contours

Fig. 4-10 Use a cardboard box *(right)* to cast a "carving brick" of plaster of paris.

ficult part of the model, clear windows, almost automatically. Paint everything but the windows on the inside of the body so that the clear plastic will provide a protective outer skin for the slot car (Fig. 4-17) or radio-controlled racer. If you're using the clear plastic body for a display model, then paint everything but the windows on the outside. The body will be very close to a scale thickness around the edges of the fender wells and around door, trunk, or hood openings to make it relatively easy to superdetail. The chassis, engine, and the interior details (including the inner panels for the doors, trunk, and hood) can be modified parts from a variety of suitable

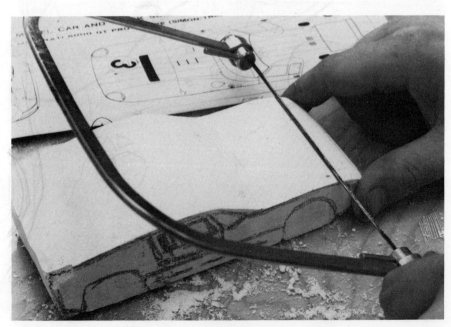

Fig. 4-11 Trace the profile of the car on the "carving brick" and use a jig saw to cut the profile shape into the plaster.

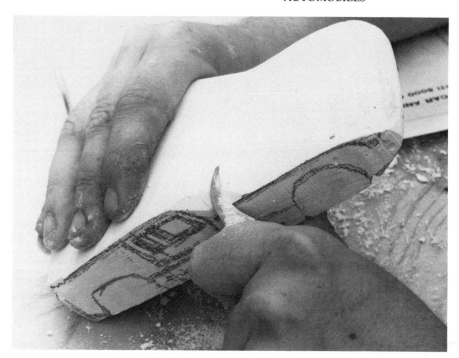

Fig. 4-12 Use a hook-shaped blade to carve the concave contours of the model, as Ron Klein is doing here.

(Fig. 4-12) and wood workers' chisels and gouges for the concave contours (Fig. 4-13). Finish the shape using ever-finer grades of sandpaper wrapped around dowels and vee-shaped blocks of wood to reach the concave areas. Add the vee grooves for door edges, window frames, and louvers as a final touch, with a vee-shaped gouge guided by a steel straight edge. When you're completely satisfied, paint the body with a two-part epoxy paint to seal it and protect it from the heat of the vacuum-forming machine. The epoxy paints are available from most automotive or industrial paint suppliers.

A vacuum-forming machine is simply a vacuum chamber with a flat upper surface that is covered by a plate (called a platen) with thousands of tiny holes. The vacuum pulls anything that is placed over the platen snugly against each of those holes. The mold or master pattern is placed on the platen so that the vacuum can pull the sheet plastic tightly

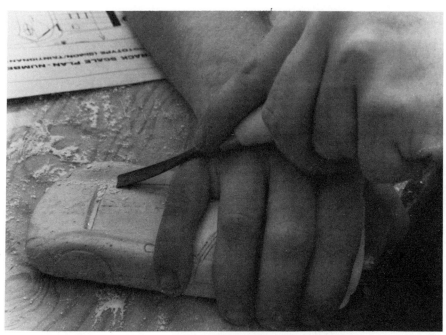

Fig. 4-13 Hollow-shaped chisels or gouges must be used to shape convex areas on the model's surface.

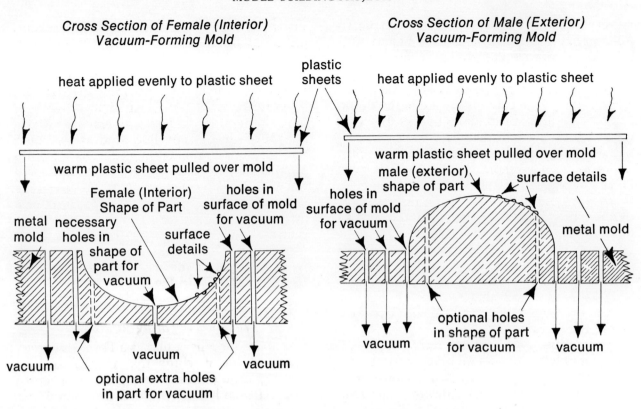

Fig. 4-14 Cross-section views of vacuum-forming molds.

around the pattern (Fig. 4-14, right view, and Fig. 4-16). Any details that are carved into the surface of the pattern will appear on the *inside* of the clear plastic. This use of the master pattern as a "male" mold works fine for most automobile bodies because most of the surface details are smooth. If you want to capture very crisp details like simulated rivet heads or louvers, these details should be molded on the *outside* of the plastic as shown in the lefthand view of Fig. 4-14 and Fig. 9-15.

You can convert your hand-carved master

Fig. 4-15 The master is supported on a platform of plaster and painted with epoxy paint with 1/32-inch holes for vacuum-forming.

pattern into a "female" mold by using a metal-filled epoxy resin such as Duro's Liquid Aluminum to capture the shape of the master. Coat a baking pan with grease to allow the epoxy resin to be removed after it has hardened, then fill the pan with just enough of the epoxy resin to cover the body. Press the body pattern into the wet resin and hold it there with some scraps of wood and weights until the resin cures and hardens. The process is far more complicated than it might sound because there is a great danger of trapping the body permanently in the epoxy resin mold. The body must be coated with grease or a special "mold release" fluid sold by industrial hardware stores. Even then, though, small indented details or undercuts may create "captures" (see Fig. 12-2) that can trap the master pattern in the mold. The pattern will be broken before the resin cracks.

If you must use a female mold, the body will probably have to be molded in two parts. I suggest you build a few of the vacuum-formed model aircraft kits (most are produced in female molds) to learn how the fuselages are molded and how they are assembled. If you use the two-part technique, you might consider making the body from white styrene plastic so it can be cemented together with an invisible joint after it is glued together with liquid cement for plastics and smoothed with automobile body filler. The windows can then be a third piece molded in either clear styrene or clear butyrate to replace the solid panels in the white styrene. You must drill holes in the female mold (and in any concave areas of a male mold) as shown in Fig. 4-14 so the vacuum will pull the plastic into these areas. Note the holes in the vents and in some of the access panel grooves in the master pattern for the Maserati 5000 GT in Fig. 4-15.

Vacuum Forming at Home

Idea Development makes a Formicator vacuum-molding machine that is intended for making large parts for flying model aircraft. The machine has a platten that is about 6¼"

Fig. 4-16 The Formicator vacuum-forming machine (left) is powered by a household vacuum cleaner (right). Clear plastic has been clamped in the frame, heated to 250°F, and pulled over a metal master in the Formicator.

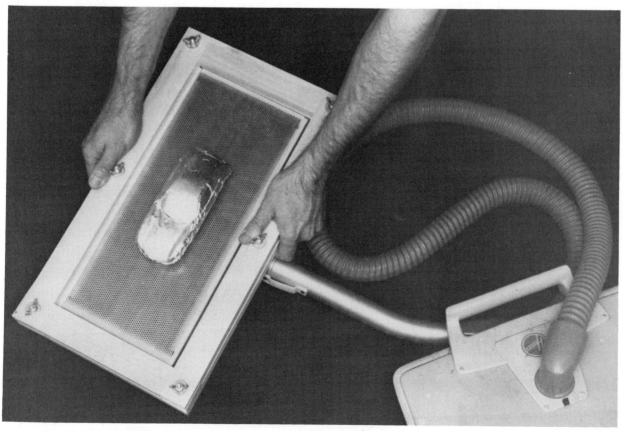

Fig. 4-17 Paint the clear butyrate body on the inside and attach it to any of the slot car chassis with screws or straight pins.

× 14½"—sufficient to form any 1/24 scale or smaller car or truck body but only about half the area needed for a 1/12 car body. The instructions furnished with the Formicator are extremely helpful in pointing out captures and other pitfalls to avoid. The Formicator does not have its own heat supply as the large commercial machines do. The tool is supposed to be placed in a home oven that has been warmed to "bake" or 250°F (with the oven door open!!). The plastic sheet must also be warmed while it is clamped in the two-piece upper frame of the Formicator. The frame holds the plastic's edges so they are not pulled inward when the frame and plastic are placed over the Formicator's vacuum chamber. Only styrene, ABS, or butyrate plastic may be used in the vacuum-forming process—other common forms of plastic sheet such as cellulose or nitrate plastics are *highly* flammable and should never be used for any model-building work. Sig and Idea Development sell the precut 8½" × 17" sheets to fit the Formicator (the larger size is needed to fit the clamping edges of the upper frame) in several thicknesses of ABS plastic. Styrene and butyrate plastics must be ordered in larger sheet sizes from a commercial plastics supplier. The .030-inch thick butyrate should be about right for most 1/24 and 1/32 scale bodies, but you might experiment with .020-inch-thick sheets for superlight bodies.

The bottom edges of the vacuum-formed body must be trimmed to size along the marks you carved into the surface of the master pattern. Most 1/32 and 1/24 scale slot car chassis have built-in mounting brackets that can be used for attaching the body. The brackets and posts that are used to mount injection-molded plastic kit bodies on the kit chassis can be used to mount the clear plastic bodies for display models. The styrene plastic brackets and details can be attached to the butyrate plastic with 5-minute epoxy.

Chapter 5
Ships

No subject for a miniature is quite as romantic as a sailing ship. This branch of the model-building hobby is, beyond doubt, the most glamorous and, historically, the most important. The sailing ship was not only one of man's first machines but also one of the first war machines. These early machines were, of course, made mostly of wood with nothing more than the wind to power them. Warships, like the *Monitor* and *Merrimac* of Civil War fame, were among the first of man's machines to be manufactured from, or clad in, iron and, later, steel. There is something that touches the soul of almost everyone in the vision of a large sailing ship breaking through the water with a gush of white water visible over the gunnels and no sound but the booming of the sails and the roar of the sea. The only machines that come as close to a harmony with nature are sailplanes, but even they are pulled aloft by machines; the sailing ship moves under nature's own power.

There is something equally awesome about metal warships that house complete cities of guns, airfields full of aircraft; ships that prowl above the seas at jetlike speeds and below them like predatory fish. Even people who consider all automobile models to be toys can appreciate a not-too-well-built sailing ship model, assembled from a plastic kit, as a "true work of the modeler's art." If the romance and glamour of ships isn't enough to inspire you to model ships as a hobby, then the appreciation of your peers for even the simplest efforts should cinch the decision for you. Everyone loves ships.

The Drydock

The range of projects for the model ship enthusiast is at least as broad as for other types of models. The range of available kits for the newcomer-to-intermediate modeler includes cast metal miniatures in 1/1200 and 1/1250 scale that are a few inches long (for use in wargames) and 4-foot-long replicas of square-rigged sailing ships with real wood hulls, metal guns and fittings and cloth sails. The most popular models, though, are the plastic kits that fall between these two extremes of size and complexity.

Some of the 1/700 and 1/720 scale plastic "waterline" series ships are relatively simple kits with most of the detail of the more modern craft molded into the decks. Some of the 3-foot-long (about 1/96 scale) plastic kits that build replicas of the famous square-rigged classics, like the Cutty Sark, Thermopylae, and U.S.S. Constitution, are almost as complex as the expensive wood and metal kits, particularly if the rigging is installed to simulate the real thing. Some experts argue that a wooden sailing ship model should have a wooden hull, but plastic can be painted so realistically that it can fool anyone. The plastic hulls on sailing ships can be particularly realistic if the decking is replaced with Northeastern's scribed basswood sheets, or with individual scale-size strips of basswood, and actual copper sheeting

MODEL-BUILDING PROJECTS

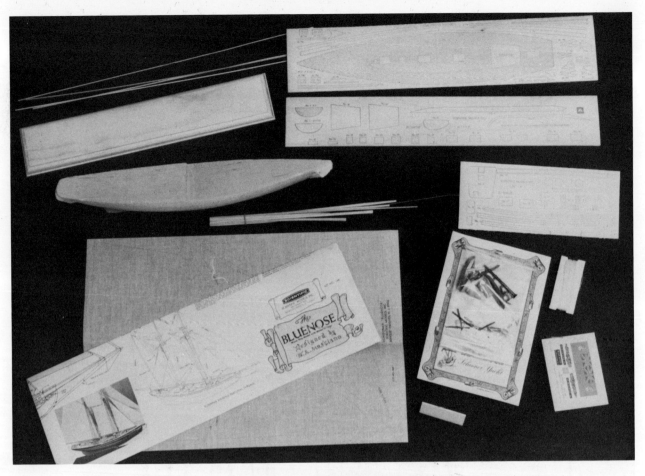

Fig. 5-1 A typical Scientific Models wooden ship kit includes a preshaped hull and display base (*upper left*), but the small parts must be cut from printed sheets of wood.

cemented over the molded-in simulated copper sheeting below the waterline. Many of the plastic hulls are quite accurate and such small deck details as hatch covers, life boats, and ships' wheels are often exquisite moldings. If you'd rather work on the spider web of rigging than whittle on a wooden hull, the plastic kits can be assembled and superdetailed to match far more expenisve kits.

With the exception of a few motorized plastic kits that have one-piece hulls, the plastic model ship kits are intended only for display. Some of the smaller scale metal ships and plastic kits do not even have a complete hull; the model stops at the waterline. These "waterline" models are usually described as such on the box lid. The flat bottoms of this type of model make it much easier to display an entire fleet (Fig. 5-3) or to maneuver the ships about in simulated sea battles. Even the wooden kits are really designed for displaying on a mantlepiece; additional lead ballast must be inserted in the keel area of the hull during the assembly stages if the boat is to remain upright when it floats.

Ships that Sail

Most of the sailing ships and replicas of modern steel warships or pleasure craft can be modified so that they actually sail (or are powered) through the water. The rigging and sails on the square-rigged ships are far too complex to be altered much after they are in place on the model. If the wind is calm and the water smooth, however, a properly ballasted model square-rigger can sail across a small lake with the wind at its stern and the rudder set to keep it running straight.

Some modelers rig a device that works like a weathervane at the top of one of the masts. An oversize replica of a flag or a piece of clear plastic is aimed downwind like a weathersock at an airport. A stiff vertical wire is attached firmly to the "flag" so the wire turns whenever the flag is moved by the wind. The wire leads below deck to the rudder through a pair of model airplane bellcranks. When the wind changes direction slightly, the weathervane/windsock/flag moves the rudder in the proper direction—after some trim adjustments by the modeler, of course—to keep the ship from heeling over too far. It doesn't take much of a change in wind direction to upset a square-rigged ship, but this device is enough to keep the ship moving on calm days.

The simple one, two, or three-masted ships and schooners can be fitted with small winches powered by electric motors. The winches in these models are controlled by a radio receiver in the hull that responds to signals from a radio transmitter in the "captain's" hands. There are radio-control transmitters and receivers with as many as seven channels. The Probar and Graupner winches are popular items with radio-controlled sailing enthusiasts. The radio-control rig's servo motors, which power the winches, can also be used on more modern ships, of course, to move gun turrets, elevate guns, launch seaplanes, or control the dive angles of scale model submarines. Relatively inexpensive "toy" radio-controlled ships can be fitted with new decks or their radio control gear used on other models (as described in Chapter 13). Remember, the ability to float upright, let alone sail under radio control, must be installed in the early stages of building the model. The batteries and radio receiver and servo motors are heavy enough to prevent their use in models with hulls shorter than about 24 inches.

Your hobby dealer can probably provide the information you need to convert a boat to radio-controlled operation. If you're in doubt, then modify one of the "toys" or build a kit, like those from Dumas, Sterling, and others, that are designed to be used with radio-control equipment. There are also a number of Fiberglas hulls that can serve as the basis for a scratch-built warship or schooner model.

Fig. 5-2 The completed Scientific "Bluenose" schooner. Courtesy of Scientific Models, Inc.

MODEL-BUILDING PROJECTS

Fig. 5-3 Mike Czibovic's collection includes hundreds of 1/700 scale waterline models made from sheets of styrene and kit parts.

Assembling Model Ship Kits

The construction principles for ship model kits are basically identical to those for rocket, automobile, or aircraft kits. The assembly and construction techniques outlined in the previous four chapters apply directly to wood, metal, or plastic ship kits, so there is no point in repeating them. There are a few special techniques that can be helpful in assembling ship models. Your first consideration must be whether or not you want to be able to actually float the model in the water or simply display it on a mantle or shelf. If you are going to float the model, you will need to add some type of ballast to the hull so the model will float at the waterline of the prototype when it is completed. It is extremely difficult to guess how much weight you must add to complete the superstructure of the model. You can certainly weigh the parts, but that won't account for cement, epoxy paint, or any details you might want to add.

The deck of the model, complete with all the superstructure fittings, should be made removable if at all possible. On some models, it is far easier to make one of the major deck details, like a deckhouse on a sailing ship or the bridge area of a warship, removable so you can reach the interior of the hull after the model is completely finished. If the opening is large enough to allow *both* of your hands to fit through, you can even install motors or fuel-burning engines to power the boat and radio receivers, servo motors, and winches for radio control. There are liable to be some "incidental" items that are easy to overlook, including such things as fuel for an internal-combustion engine or batteries for an electric motor and batteries for any radio receiver.

The hull should be sealed (so it will float) with automobile body putty along any seams you suspect might leak. Wise modelers fill every available nook and cranny of the hull with chunks of the white Styrofoam plastic that is used to pack small electrical parts and large refrigerators for shipment. The expanded foam plastic chunks are sometimes enough to keep the ship afloat even if it capsizes or the hull springs a leak.

Superdetailed Rigging

Almost every sailing ship kit includes some deliberate simplifications of the rope rigging to make the kit a bit less formidable to complete. It takes an experienced seaman or ship modeler to spot errors in rigging. Even a novice can notice when the lines are far too thick or too uniform in size, and that's a mistake that exists in almost every kit. Scale-size linen lines are available from firms like Model Shipways to allow a variety of different-size lines to match the variations on full-size ships. This one superdetail can be enough to make even an inexpensive plastic kit look like a handcrafted wooden ship.

Fred Turneir is a professional ship modeler who builds special miniatures from scratch for museums and rebuilds centuries-old historical models. Fred has carved a vee-notch in the end of an 18-inch length of 3/16-inch diameter dowel. The notch is used to guide the lines (linen thread) around the belaying pins, blocks, and other tie-off points on the ship.

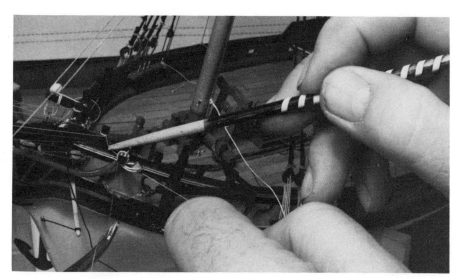

Fig. 5-4 Fred Turneir made this knot-tying tool by carving a small yoke in the tip of a hardwood dowel. The stripes make it easier to find on the workbench.

He uses a pair of self-locking forceps to help hold the free ends of the lines (Fig. 5-4). He superdetails some of the models he custom builds from plastic kits using real copper hull sheeting (.002-inch thick), chain, brass stanchions, and pins from Model Shipways. He also hand-shapes real wood dowels for the masts and spars.

Ship Kit Conversions

When you have completed one or two out-of-the-box ship kits with, perhaps, some superdetails such as linen thread rigging, you may want to make something you cannot buy in any kit. There are two ways to create such a "one-of" model; you can build it from scratch using scale plans, or you can modify a kit to produce a ship that is different from that intended by the kit manufacturer. The process of altering the kit is called a conversion by most modelers, because it requires that the kit parts be "converted" by altering their shape or size. Some conversions require the use of parts from several kits. You might even consider using a different paint scheme or relettering of, for example, a destroyer to match a different member of the same class to be a conversion. The fundamental principle behind any conversion is research; you must know enough about both the stock kit or kits and your proposed vessel to be able to provide all the details. Even a simple name-change to another ship in the same class of warships requires the research to determine if all members of that class really were identical.

There is plenty of resource material available on every imaginable type of ship or boat from any period in history. Your local public library should be one of the first places to look because that information is free. The firms listed in the back of this book that supply plans often have catalogs of additional publications with photographs and drawings.

The plans shown on these pages (Figs. 5-5, 5-6, and 5-7) are examples of some of the Alan Chesley drawings sold through the Floating Dry Dock company. When you purchase this particular series of plans, they are about the size published here: ⅛ inch = 1 foot. If we had published them in the popular 1/720 scale used for many waterline models, the details would have been completely invisible. You can use these plans and other prototype information and photographs to verify the possibilities of creating a kit conversion from one of the existing waterline kits or, if you must scratchbuild, the hull and major deck structures. If you stick to a scale with a number of available kits (like 1/700 or 1/720), you can use the turrets, guns, and other small details that are so time consuming to carve. Fred Turneir is working on a simple conversion of the popular Model Shipways' "Essex" square-rigger; he is converting that wooden kit into the almost identical "Raleigh" by shortening the hull about 9 scale feet while he assembles the kit. The rigging, of course, will have to be matched to the plans and photographs of the "Raleigh" to complete the conversion.

Scratchbuilding Waterline Models

There are really no plans published expressly for the use of modelers who like to build in the "collector's" or wargamer's scales of 1/700 and 1/720. There are dozens of plastic model kits for ships in these very similar scales, from firms in America, England, and Japan, with replicas of ships from those three nations' World War II-era navies. There are just enough "missing" models to make it worthwhile to create your own to duplicate a complete battle scene or even an entire fleet. The 1/16-inch = 1-foot plans from the Floating Dry Dock and from Edward H. Wiwesser can be used full-size for shaping and positioning details with photostatic reductions (as described in Chapter 1) to either 1/700 or 1/720 scale. The plans in Figs. 5-5 and 5-6 are almost twice the size needed for 1/700 or 1/720 scale models, but we used them to demonstrate how you can scratchbuild the hulls for waterline models using .020-inch thick styrene plastic sheets. The plans should be photocopied so you can work with them on paper that is blank on one side. Place the plans on a piece of ¼-inch-thick clear Plexiglas (most glass shops sell it); then place the .020-inch-thick white styrene plastic over the plans. Hold the sandwich of Plexiglas, styrene, and plans over a light and you'll be able to see the plans right through the plastic (Fig. 5-8). You can then trace the profile of the bottom of the hull, the sides of the hull, and at least three bulkheads on the .020-inch white styrene plastic. Use the score-and-break technique (shown in Chapter 9) to snap the parts from the sheet of plastic.

Tape a piece of waxed paper to a perfectly

flat scrap of 1" × 4" wood and use that as the assembly table for the hull. Glue the bulkheads to the bottom of the boat using liquid cement for plastics. Next, join the sides of the boat at the bow, using a clothespin or hairpin to hold them together for about 24 hours. The sides of the hull can then be cemented to the bottom of the hull and to the bulkheads (Fig. 5-9). Allow 48 hours for those glued joints to dry before wrapping the sides around the curved stern of the boat.

You will have to judge the proper length for the sides by measuring their curved length from the *top* view of the plans—the deck's curved length will have to be measured from the side or profile view of the plans. A small reinforcing strip of styrene will be needed to reinforce the joint between the sides where they are joined at the stern. The radiused edges of the bow and the stern may require some additional material to build up their shape. Apply a thin layer of automobile body filler (spot) putty to the bow and the stern and let the putty and the glue joints dry for 48 hours. The bow and stern can then be sanded to the precisely correct shape using fine sandpaper glued to a short piece of ¼-inch diameter wood dowel to reach concave areas.

Install the deck next. The larger deck houses and cabins can be fabricated from strips of .010 and .020-inch-thick white styrene plastic as hollow blocks. The round-cornered fittings, such as some gun turrets and most smokestacks, are easiest to shape from solid blocks of styrene. The blocks can be made by simply laminating several pieces of the thickest styrene you have on hand. Sheets up to .080 inch thick are relatively easy to obtain from hobby shops that carry Evergreen styrene or Plastruct ABS plastic. Use liquid cement for plastics to bond the sheets together and let them dry for at least 48 hours before sanding or filing them to match the shapes of the smokestacks or other details. Plastruct offers a variety of telescoping sizes of plastic tube that can be used for guns or flattened into ovals for some types of stacks on early war ships. The K & S brand brass tubing or even hypodermic syringe needles can be used for smaller guns.

Shaping Wooden Hulls

There is something very satisfying about shaping the hull of a model sailing vessel. The process is made fairly easy thanks to a wide number of hull plans with profile sections of both the vertical bulkheads and the horizontal waterlines. These sections can be traced onto pieces of cardboard and cut out to be used as test gauges to make certain the shape of your miniature hull is an exact match for the hull on the full-size ship. The sources of ship model plans in the back of this book can provide precisely the plans you need for the hull (and most other details) of almost any known sailing vessel from any era. The plans will most likely be in a scale or size that does not suit your model, but it's simple enough to have them photostatically reduced (or enlarged) to the exact size you need. The solid wood hulls in most of the ship kits are preshaped so that only a few light cuts and some light sanding will be needed to finish them. The size of solid wood hulls can be reduced slightly if their shape is incorrect, but there's nothing really suitable for enlarging any areas that are too small. The kits with hollow or planked hulls, however, can be corrected by simply reshaping the horizontal ribs (sometimes called bulkheads or frames). The ribs can be shaped to correct errors in the shape of the hull (in either a horizontal or a vertical plane) by matching each of the ribs to the sections from the plans of the full-size vessel.

If you are building your ship model from scratch, you have the same choices of hull construction available to the manufactures of kits: solid wood or the hollow "planked hull." The only portion of the hull that will be visible when the top deck is in place is ordinarily that of the bulwarks that extend upward from the deck. Strips of wood or "planks" are provided for the bulwarks in most kits, and you can follow that same type of construction for your own models. Some modelers prefer the hollow hull construction because they find it easier (and more realistic) to build up a hull than to reduce a square block to the shape of a hull. Having the hollow hull is a definite advantage if any portions of the lower decks are to be visible through gunports or transoms. You can, of course, carve square holes in a solid wood hull for the gunports using a milling bit in a motor tool like Dremel's (see Fig. 5-13). If the backs and sides of these holes are painted dark brown and the holes filled with cannon muzzles, the effect is quite convincing. I suggest you try a kit with a solid hull (from Sterling or Scientific) and one with a hollow hull (from Sergal or Billings) to see which method of construction is most enjoyable.

MODEL-BUILDING PROJECTS

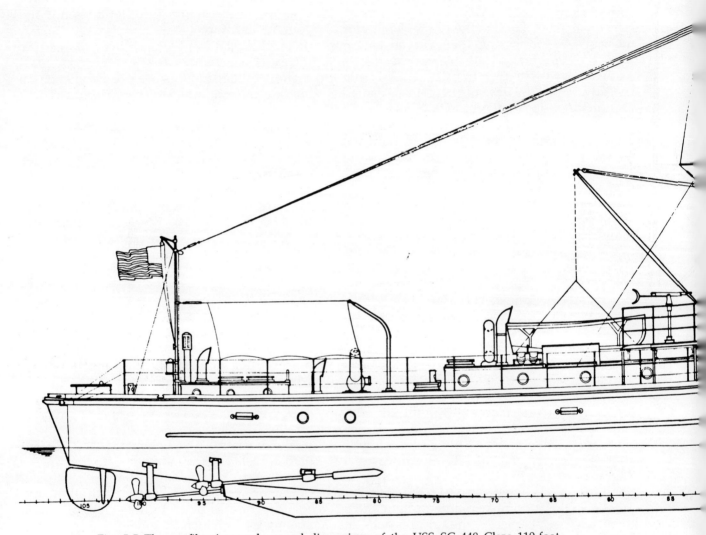

Fig. 5-5 The profile view and general dimensions of the USS SC 448 Class 110-foot Submarine Chaser (circa December 1948). This tiny ship would make a fine project for a 1/400 scale waterline model (about 3-inches long); in the more popular 1/700 scale, the model would be a scant 1.88 inches long. Plans courtesy of the Floating Drydock.

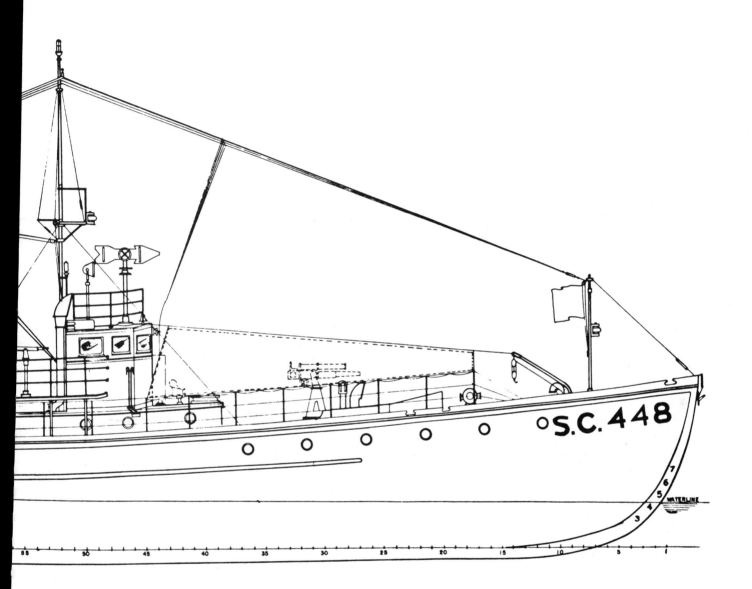

MODEL-BUILDING PROJECTS

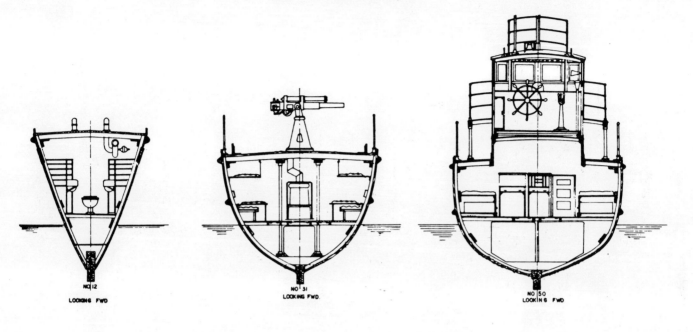

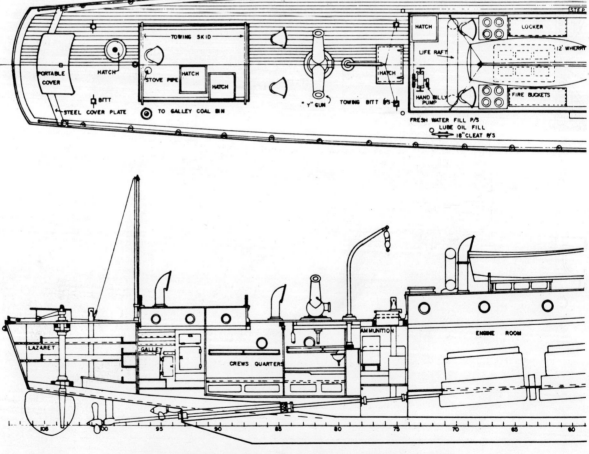

Fig. 5-6 Cross sections of the hull (bulkheads) provide the shape of the rather complex ship. Each cross section is taken from a specified distance (in scale feet) from the bow of the ship. The lower view is a cutaway from the side to show additional deck and interior details. Plans courtesy of the Floating Drydock.

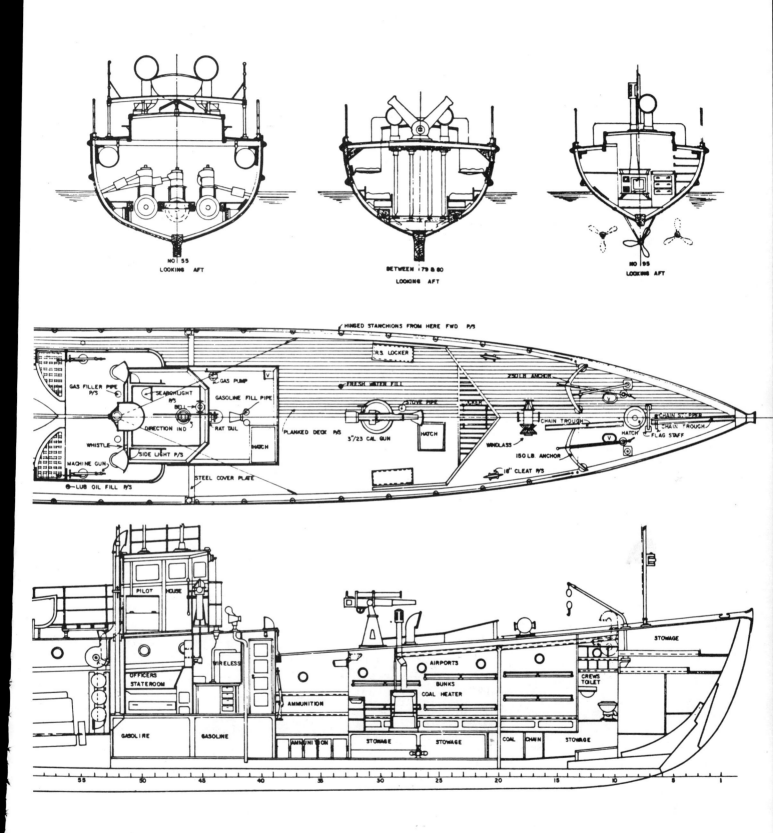

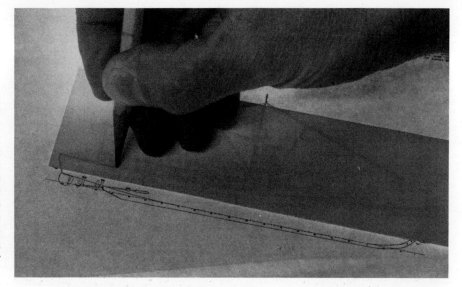

Fig. 5-7 A light held beneath the plans, which are held on a sheet of clear Plexiglas, will allow you to trace the plans precisely on a sheet of 0.15 or .020-inch-thick white styrene plastic. If you are working with full-hull plans (like this Submarine Chaser), be sure to make the bottom of the ship at the waterline marked on the plans. The length of each side must match the length shown on the top view to compensate for the curvature of the side.

Finishing the Hull

Solid-wood hulls can be covered with individual planks so they appear, from the outside, as realistic as a hollow hull. If you are going to carve the hull from a solid block (or a lamination of several solid blocks), then you can use a relatively soft wood like hard balsa or obechi. If you intend to simply scribe or ink the joints between the planks, then you'll want a fine-grain wood like walnut or the rare jelutong wood from China. Walnut plywood of 1/8 inch is probably the best choice for the ribs or frames and keels. The plywood must, of course, be cut with a hand or electric jig saw or an electric saber saw and the edges sanded smooth to match the hull plans. The planking, whether for a hollow hull or a carved hull, can be matched to the type of wood used on the full-size ship if you can find a cabinet shop willing to cut the strips you need. If you must use precut wood, then basswood is the most readily available material (from Northeastern, Kappler, or Midwest). Midwest also offers precut strips and sheets of cherry, walnut, pine, maple, and mahogany in sizes suitable for ship models.

The strips must be pinned to a solid hull until the glue dries, but only the first two or three strips need be pinned to a hollow hull; the others can be held with clothespins (Fig. 5-14) to the strips above them. White glue,

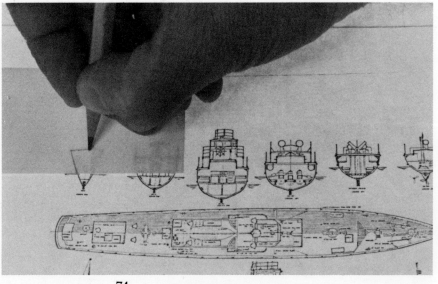

Fig. 5-8 Trace the cross-sectional views of the hull to serve as bulkhead formers or cross braces. For a small-scale (1/400 or 1/700) waterline model, you can simply trace a straight (rather than curved) line for the vertical edges of the hull where the sides will attach to the cross braces. Use .040 to .080-inch-thick white styrene plastic for the cross braces.

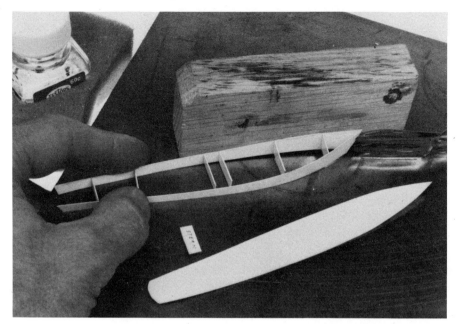

Fig. 5-9 Liquid cement is used to glue the cross braces (bulkheads) to one of the ship's sides. When glue is dry, cement bow of the second side to bow of the first and hold with metal clip. The separate stern piece will be installed after the deck, so the curve can be used as a guide for the stern's shape. Use .020 to .040-inch white styrene to build the bridge and other structures on deck.

plastic resin-based glues (like Titebond), or Goldberg's Super Jet thickened cyanoacrylate can be used depending on what you can afford and how quickly you want the glue to dry. White glue and Titebond dry in about 12 hours; Super Jet dries in about 5 minutes. It's much easier to make the plywood keel, on a hollow hull, extend most of the way upward toward the top deck with notches in the bulkheads fitted to matching notches in the keel

Fig. 5-10 Mike Czibovic uses both styrene plastic (*white*) and Plastruct's ABS plastic (*gray*) for his scratch-built 1/700 scale waterline models.

Fig. 5-11 A 1/700 scale is just large enough to allow the use of such tiny details as guns and aircraft from the plastic ship kits.

(see Fig. 5-14). X-Acto's aluminum C-clamps can be used to make the keel and bulkheads a self-standing unit while you add the exterior planking, the decks, and the other details.

Most of the sailing ships that are popular with modelers were fitted with copper plates on their hulls below the waterline. These plates would soon weather to dull gray-green when the ship was in use, so you may see color paintings or photographs with the lower hull green rather than copper-colored. You can decide whether you want your model to have

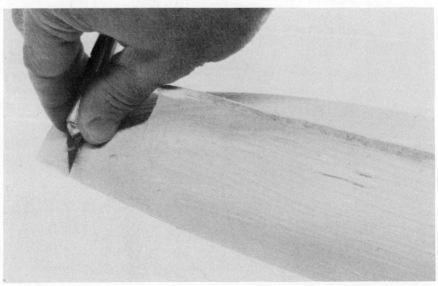

Fig. 5-12 The wooden hulls in ship kits, such as Scientific's, are preshaped but the stern must be carved away and the surfaces sanded smooth.

SHIPS

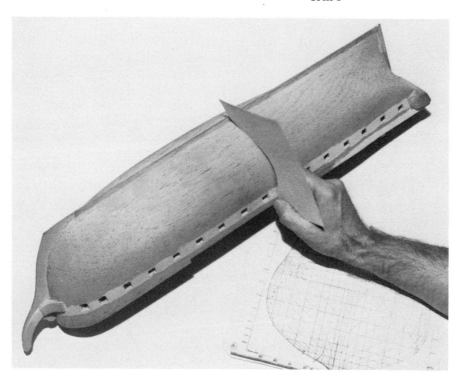

Fig. 5-13 The plans for a ship's hull include cross-sectional views all along the hull. Make cardboard patterns to be sure you are carving the hull's shape correctly on *both* the right and left.

Fig. 5-14 Fred Turneir uses a plywood central bulkhead/keel piece with plywood bulkheads for his individual-plank hollow hulls.

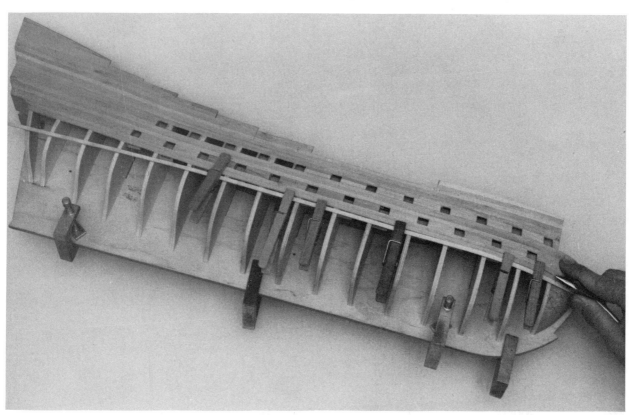

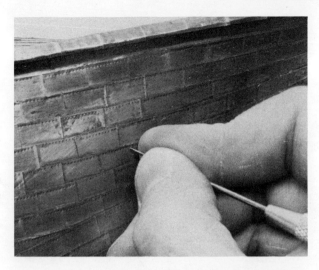

Fig. 5-15 Model Shipways' .002-inch thick scale-size copper sheeting is the perfect material for most sailing ships' hulls.

copper-colored plates, as though it just left the drydock, or the greenish tint of a ship that has just been pulled from the water.

The Model Shipyard offers several sizes of .002-inch thick copper plates cut to size. These plates can be installed on almost any type of model. The copper is thin enough that you can burnish the plates over a plastic kit's hull so the kit's simulated nail heads will show through the .002-inch copper. If you are applying the copper plates to a wooden hull, you'll have to make your own nailhead marks. The nail heads generally left a slight dimple which is more visible than the actual head of the nail. You can simulate both the nail head and its indentation by simply pushing a compass point or scribe into the plates after they have been attached to the model hull (see Fig. 5-15). Use contact cement to hold the plates to a wooden hull. The contact cement will melt a plastic hull, so the plates must be attached with 5-minute epoxy or Goldberg's Super Jet. The patterns that were used to apply the plates and to protect the keel and rudder were distinctive features of many ships, so be sure to follow the plans or photographs of the prototype carefully. When the hull is completed, protect the copper color with a thin spray-on coating of Testors' Dullcote or Glosscote.

When the exposed wood portions of the hull and the masts and spars are painted, the grain will rise enough that the parts will look quite crude. At least one coat of sanding sealer should be applied to most woods before a final coat or so of shellac or Walthers' DDV flat varnish. The shellac or varnish can be rubbed down with fine steel wool or No. 200 wet-or-dry sandpaper if any of the grain is still too harshly defined. Scale model ships, even the 4-foot-long models, look much more realistic with a flat or semigloss finish than with a glossy one. If you do use glossy paints to color the model, cover them with a dulling coat of Walthers' DDV or Testors' Dullcote.

If the model is actually going to be used in the water, the wood should be sealed (both inside and out on any hollow hulls) with at least four coats of shellac or DDV. Apply a fresh coat of wax to protect the hull below, and immediately above, the waterline just before the model is to be launched into the water.

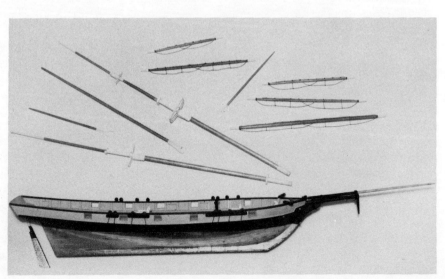

Fig. 5-16 The hull, masts, and spars for Fred Turneir's Model Shipways kit-built craft are ready to be installed and rigged.

Chapter 6
Military Vehicles

MAN'S genius for creating machines to fill his needs has been most evident in the vehicles he designs for war. Today, we call these "war" machines "defense weapons," but the leading edge of technology continues to center on engineering works that are at least equally well suited to war and to peacetime uses. Whether we find the history of warfare fascinating or appalling, most of us must admit that the machines that waged war are some of the most incredible creations in history. There are almost as many books on the vehicles used in the various wars since the twenties as there are books on aircraft or automobiles. Those books illustrate the various military vehicles that have existed in photographs and, sometimes, with color drawings and plans or profile views.

There is no better way to depict any machine, however, than to build a three-dimensional replica of it to an exact scale. Enthusiasts lump all military vehicles under the heading "armor," armored fighting vehicles, or AFV models. Those who prefer to paint the extremely detailed tin and lead soldier figures consider their hobby "military miniatures." There are dozens of plastic soldiers included with many of the armor kits and "accessory" packs of soldiers, simulated barbed wire, broken walls and fences, and military supplies (like fuel cans, ammunition crates, and spare parts) are available in plastic to match the scale sizes of the armor kits. These items are intended to make small dioramas with scenery to display armor models (Fig. 6-1). Many of the metal soldiers are made in sizes to match the popular plastic armor kits (or vise versa) so the hobbies of modeling the military vehicles (armor) and the soldiers that used them (military miniatures) are compatible.

AFV Miniatures

Nearly all the kits for armored fighting vehicles are injection molded in plastic. There are a few 1/250 and 1/285 scale metal models, but these are intended primarily as playing pieces for war games. There are also a few vacuum-formed conversion kits to be used to make a totally different vehicle from some existing injection-molded kit. There is little need for vacuum-formed kits, however, because the flat surfaces of most armored vehicles make it relatively easy to create your own conversions, like those described in this chapter.

The model building techniques described in Chapter 2 and the painting techniques outlined in Chapter 1 apply to all the various brands of armored fighting vehicles. Most of the vehicles classified as AFVs are tanks, half-tracks (armored machines with wheels for steering and tracks on the rear for traction), and various types of mobile or stationary artillery. The unusual four or six-wheeled armored scouting vehicles and various types of armored personnel carriers are certainly included in the AFV category. Any machine used by any of the armed forces falls into this category, however, including "soft-skinned" (nonarmored) vehicles such as motorcycles, jeeps, ambulances, and automobiles used to transport officers.

An extensive range of 1/87 scale (HO train-size) ready-built AFVs from Roco are sold under the Mini-Tanks name. There are also several manufacturers of 1/76 scale plastic AFV kits (British 00 train-size) including the Airfix line from England and the Japanese Fujimi series. ESCI 1/72 scale kits are imported by ScaleCraft, but the complementary range of Hasegawa 1/72 scale AFVs must be ordered from one of the Japanese hobby shops because they are not imported as part of the American Minicraft/Hasegawa series. Airfix of England and Monogram also make a few vehicles in 1/32 scale, but 1/35 is the most common "mid-size" scale for armor miniatures. Tamiya's models are imported by MRC, Testors is producing the Italeri series, and the French Heller line is available through Polks. There are also a few 1/48 scale kits that occasionally appear and disappear from the market as well as some motorized and radio-controlled tanks in the larger 1/24 and 1/16 scales.

If you are more interested in collecting than in building or superdetailing, I suggest you stick to 1/87, 1/72, or 1/76 scale armored models. If you enjoy adding extra details or building conversions, then 1/35 scale should satisfy your needs. Few modelers have the space for more than a few tanks or other AFV miniatures in the 1/24 or larger scales.

Weathering

Vehicles in the AFV class were designed to live in the dirt. Most of the full-size machines were delivered from the production lines with flat-finish paints and camouflage patterns that were meant to blend into the terrain. When these vehicles see action, the dirt-like look of their paint becomes partially covered with dust, mud, and scratches. Tanks, half-tracks, and even the soft-skinned machines soon acquire a very "used" appearance. The simulation of that effect on a miniature is called "weathering."

There is some disagreement among modelers about the use of weathering even on an AFV; one school prefers to build and paint their models so they look like new and the other school feels that a miniature should capture the most common appearance of the prototype, which means that at least a touch of "road dust" is applied to all of their models—most are much "dirtier." This same argument prevails among model railroaders and aircraft modelers for much the same reasons. Most automobile and ship modelers are content to finish their miniatures in an "as-built" condition.

Fig. 6-1 Gary Maisak added some 1/35 scale MRC/Tamiya soldiers and a facial tissue top to this 1/32 scale Airfix (England) Humber and mounted the model on a diorama with scenery.

The Art of Weathering

It is far more difficult to capture the appearance of weathering on a model than you might suppose. A sloppy paint job on a model will look like a sloppy paint job even if it is supposed to represent splashed mud or chipped and peeling paint. You must master the techniques of painting miniatures in their as-built state before you advance to the stage where you can give your miniatures a weathered appearance. When you paint a model, your work is about 90 percent technique and 10 percent "art"; most of the skill involves getting the paint on in the thinnest possible coat with proper color separation lines and smooth edges around windows. Weathering is about 90 percent art and 10 percent technique; it takes a considerable amount of practice to be able to weather a model convincingly.

The weathering starts with the first coat of paint on the raw plastic. If you want to simulate peeled paint with bare aluminum or steel showing beneath the paint, then the first coat of paint should be an aluminum or steel color. When the paint is dry, brush a thin layer of rubber cement (sold by stationary stores) or artist's matte medium (sold by art supply stores) in the areas where you want the color coats to "peel." When you apply the color coat, brush (or, better, airbrush) the paint over the entire model. When the paint is dry, you can pick it and peel it back on those precoated spots to have truly peeled paint effects. If you want the paint to appear faded by the sunlight, add white to the colors. It takes a considerable amount of white to make much difference; for really faded shades it's easier to add color to the white.

Most of the weathering effects that can be applied after the model is painted are actually attempts at simulating some types of damage or overcoating. The top surfaces of the full-size vehicle will often appear to be a lighter color because the paint has faded from exposure to the sun or the paint may have been worn away by wind or sand. Those effects are simulated by mixing white into the paint itself. The lower edges of the vehicle will also appear faded, but that may be caused by an overlayer of mud or sand. The mud or sand may also have worn some of the paint away, but the wear will be in the form of pebble-size chips rather than a general fading.

If you painted the model with a "primer" coat of aluminum, the chips can be simulated by merely picking lightly at the surface with the tip of a hobby knife. The knife should barely touch the plastic and, when it is flicked aside, a real chip will appear that is surrounded by the aluminum undercoat. If the plastic color shows, touch it up with a two-hair brush load of the original aluminum paint.

Mud can be simulated by just touching the very tip of a paint brush into a bottle of beige or brown paint. Dab most of the paint off on a scrap of paper and touch the tip of the brush to the model so that only a few dozen dots of paint appear where the tips of the bristles touched. An airbrush is extremely helpful in simulating the layers of dust or stains from exhausts, smokestacks, or gunpowder burns. Use thin dark brown or dark gray paint (not black paint) with about nine parts thinner to one part paint. Adjust the airbrush so that only a small pattern of paint appears with the air pressure set at 25 to 30 pounds per square inch. Hold the brush about 12 to 18 inches from the model so that the paint will literally be dusted onto the surface. On some types of vehicles, this dust may have been washed down the sides by rain; simulate this effect by carefully streaking the model with a No. 00-size paint brush dipped in thinner. The recently applied "dust" of thinner and beige can be streaked to look exactly like rain-streaked stains.

Keeping It Subtle

Most new modelers make the mistake of overdoing any weathering effects that simulate rust or worn paint. The paints labeled "Rust" are almost always too orange; rust is a slightly reddish brown, similar to what is labeled "Boxcar Red" or "Roof Brown" in model railroad paint colors. Rust stains should be applied by streaking the model with a mixture of nine parts thinner to one part reddish brown.

Don't attempt a rust stain unless you have a photograph of a similar prototype with a rust stain so you can judge the position and shape of the stain. Most rust appears in the form of irregular-shaped tiny dots where the paint has chipped, allowing the iron or steel beneath to rust. Apply such dots with a two-bristle brush using the reddish-brown paint right from the bottle. It's common to see chipped paint near rust spots or even beside them, so you might want to use the techniques in the previous paragraph as well as "rust." The combination of the two techniques is especially effective on the treads of a tank or other tracked vehicle. Be careful, though, because some full-size tanks had rubber inserts in their tracks for paved road travel.

The aluminum-colored Rub'N Buff paste sold in craft supply stores is popular with some modelers. Just a trace of this paste can be used to highlight wind-worn and exposed rivet heads and to accent the wear on tracked vehicles. You can achieve a far better effect by prepainting the model with aluminum followed by the color coat; when the color is dry, rub it with No. 600 wet-or-dry sandpaper or fine steel wood so the aluminum primer shows through just the way it does on the actual vehicles.

Armored Conversions

Full-size AFV machines are often just variations on a basic chassis design. The pressure of wartime production and the high cost of manufacturing force most countries to apply many different superstructures or bodies to just a few different chassis. This makes the task of creating a wide variety of AFV models much easier for the modeler because the chassis he or she needs is probably available in a kit.

The 1/35 scale Sturminfanterie Greshetz 33 (StuIG 33) version of the World War II-era German Panzer MkIII tank is an example of a relatively simple conversion. Dan Wilson started with a Tamiya 1/35 scale Panzer MkIII kit and fabricated the large armored enclosure for the cannon from .020-inch-thick sheets of styrene plastic. The conversion merely fits over the stock kit chassis and hull; the only differences are extra fittings like spare wheels, tarps, and fuel cans. This same conversion can be created using a 1/48 scale Bandai brand Panzer MkIII or the 1/72 scale ESCI version of the same tank. The plans for the StuIG 33 parts will have to be photostated to a smaller size, of course, to match the 1/48 or 1/72 scale kits.

The score-and-break techniques of cutting

Fig. 6-2 Dan Wilson's out-of-the-box MRC/Tamiya Panzer MkIII *(left)* and the conversion to the StuIG 33 *(right)*.

plastic described in Chapter 9 were used to make the outlines of the parts. The square holes were gouged with a hobby knife and filed to precise size with a flat jeweler's file. Dan used small pins, inserted into holes drilled with a pin vise, to simulate the 1/35 scale rivets. Rivets for smaller scale models can be embossed as shown in Chapter 9. The gun barrel was sanded to a taper from a length of Plastruct ABS plastic tubing.

To build the model, assemble the chassis and hull from the kit. Cut all of the parts to shape using the "see-through" tracing techniques (Fig. 6-7). Test-fit all the parts and file or sand any edge that does not align properly. Apply the rivets and the strips of plastic that detail the front panel. The hatches are two layers of plastic, one the exact size of the hole and the other .020-inches larger to form the upper surface of the hatch. Assemble the panels using liquid cement for plastics and cement them to the hull. When the cement is dry, fill any gaps with automobile body filler (spot) putty and file the putty flush with the surface after it dries. The model can then be painted. It is almost impossible to duplicate the camouflage scheme Dan used on his model without the use of an airbrush (see Chapter 1).

Superdetailing Models

The StuIG 33 features a number of super or extra details that can be used on many other models to enhance their realism. Dan bent some .015-inch diameter steel piano wire to form the handrails and radio antennae. The wires are inserted in holes drilled into the plastic after the model was complete. Extra fuel cans, sections of track, and tires are included in some kits or you may find them in "accessory" kits in the appropriate scale. Rolled-up tarps can be simulated by careful cutting and folding facial tissue into the proper shape and then spraying it with flat-finish paint. Surgical gauze is about the right size for camouflage netting when cut into strips and rolled to match the netting carried on some full-size AFVs. Beige nylon sewing thread can simulate the ropes that tie tarps and nets to the tops of vehicles like the one illustrated here.

Combining Scratchbuilt and Kit Parts

There are enough kits on the market to build just about every popular armored or soft-skin vehicle of both the past and the present. If you search carefully, you should be able to find a kit, in the scale you prefer, that can supply the wheels, tires, tracks, guns, and suspension for just about any scratchbuilding project. You could certainly fabricate these parts from strips and disks of plastic, but there is little reward in simply duplicating what you can buy. There is a considerable thrill awaiting you, though, when you complete your first mostly scratchbuilt AFV.

The hull or superstructure is the dominant feature of any machine, and this is the part of the project you can fabricate. The flat armored plate panels can be duplicated easily enough by cutting them (using the labor-saving score-and-break technique) from styrene plastic sheet. You could certainly build one of these models from metal or cardboard, but the finished product would not be any more realistic than if you used plastic. Plastic is the easiest and quickest of any model-building material to join at any corners, and it lends itself nicely to the use of the tracks, wheels, guns, and details from existing plastic kits. Many of the books and magazines that deal with both miniature and full-size AFVs include plans that can be copied and photostated to the correct size for your personal models.

Building a Japanese Half-Track

No plan existed for the Japanese half-track "Ho-Ha" Type 1, so Dan Wilson used photographs in the *Observer's Fighting Vehicles Directory, W. W. II* by Bart Vanderveen (Warne, London, 1969) and Harold Johnson's *Tank Data 3* (WE, Inc., Old Greenwich, CT, 1972) to draw his own scale plans (Fig. 6-6).

The Japanese built the majority of their World War II-era armored fighting vehicles on several variations of the same track system. The Japanese "Chi-Ha" tank uses the same track and suspension as the "Ho-Ha"; the kit is available in 1/76 scale from Fujimi and Airfix and in 1/35 scale from MRC/Tamiya. Aurora once offered the model in 1/48 scale, and Monogram now has the dies so they may offer it under their own label.

The front axle of the "Ho-Ha" Type 1 half-track was not driven, but it did utilize a wheel that is almost identical to the current American highway truck wheels rather than the bolt-together style that was common during the forties. The front tires can, therefore, be taken from almost any half-track kit in the proper scale. AMT offers 1/43 scale truck kits that

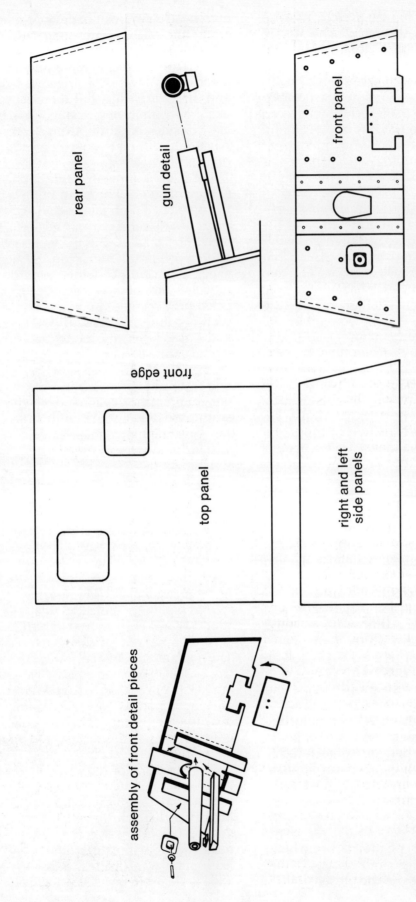

Fig. 6-3 Full-size plans for the superstructure of a 1/35 scale StuIG 33 version of the Panzar Mark III German tank, with an exploded view of the gun and front plate details (*left*). Courtesy of Dan Wilson and Miniature World.

MILITARY VEHICLES

Fig. 6-4 The rivets on a 1/35 scale model can be duplicated by simply inserting straight pins into holes drilled in the plastic.

could supply the front wheels for a 1/48 scale model; to make a 1/35 or 1/72 scale model, a track guide wheel (bogie wheel) from a 1/24 scale (for a 1/35 scale model) or a 1/48 scale (for a 1/72 scale model) tank kit must be trimmed down and eight holes drilled to match the pattern on the plans. The front axle can be fabricated from several scraps of molding sprue.

Making Plastic Armor Plate

A single 6½" × 10½" sheet of Evergreen white styrene plastic should supply all the material needed for a Japanese half-track. Use .020-inch-thick styrene sheet for 1/72 or 1/76 scale versions of Dan Wilson's miniature, .030-inch-thick sheet for a 1/48 scale version, or .040-inch-thick sheet for a 1/35 scale version. Make a photocopy of the plans (or of your photostatic reduction of them) and place that copy over a sheet of ¼-inch Plexiglas. Hold the Plexiglas and plans over a light and cover the plans with your sheet of styrene. You can then trace the plan's panels directly onto the styrene (Fig. 6-7) using a steel ruler to guide your pencil. The angles on the various panels may create some confusion if you are not careful—note that the front bumper and the tailgate are the *only* two panels that appear on the plan in the exact scale size; all other panels are angled away from the plane of the drawing. This gives the panels some perspective, which foreshortens their length. To make patterns for the panels, then, you must use the draftsman's technique of "developing" or "generating" the correct-length lines and shapes.

Let's use the upper side panels (Fig. 6-8) as an example. The side panels' *length* is shown in the side view, so measure the length of the top and bottom lines of the side panels and mark them on the sheet of plastic. Use a drafting triangle or a machinist's square (see Fig. 9-1) to be sure that all the lines you draw are straight, that all right-angle corners really are 90 degrees, and that lines that are supposed to be parallel really are. A compass (Fig. 6-8) will help to check the lengths of some lines and to double check that other pairs of lines really are parallel. The height of this upper side panel is shown only on the end views of the plan (Fig. 6-6). The three angles at the front edges of this panel are *not* shown anywhere on the plan; you must draw these after you have determined the length and height of each of the corners where these angles meet from the side and end views. Use the same method to develop the shape of the right and left lower side panels.

It is not necessary to develop the shape for every panel on the model; use the score-and-break technique to snap the two upper side panels, the two lower side panels, and the rear panel from the sheet styrene plastic. Make a second end panel to match the first, but in-

Fig. 6-5 Drill holes in the plastic to fit .015-inch steel music wire for the radio aerial and grab irons.

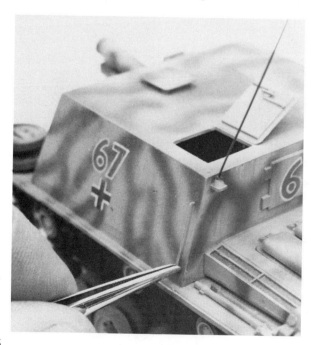

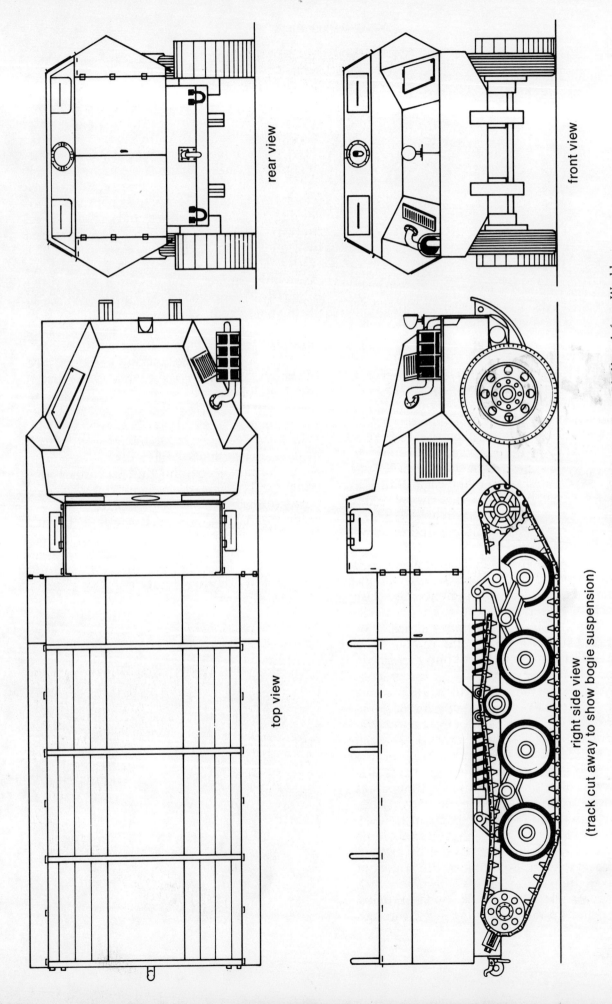

Fig. 6-6 Full-size plans for a 1/35-scale Japanese "Ho-Ha" Type 1 half-track from World War II. Courtesy of Dan Wilson and Miniature World.

MILITARY VEHICLES

Fig. 6-7 The vertical panels can be traced directly onto the sheet plastic using a ruler as a guide.

crease its height (as shown on the rear view of Fig. 6-6) so it can serve as the back panel for the cab. This same pattern must be used to shape the three interior braces (Fig. 6-10) and .040-inch round (for a 1/35 scale model) frames that hold a canvas cover over the rear of the half-track's body. You must remember to subtract the thickness of the plastic of adjoining pieces when you make each part.

It will be easier to build this model if the side panels overlap both the rear end panel and the back of the cab panel; if you're using .040-inch thick plastic, then you must remove .040-inch strips from *both* sides of the rear end, the three interior braces, and the cab panels. When you have some experience building from sheet plastic, you can plan ahead so you'll know which parts to make shorter; for now, make them full-size and remove .040 inch whenever you need to fit one part inside another.

Use the score-and-break technique to remove the side doors from the upper and lower side panels. The three interior braces and the rear end panel can now be cemented to the upper and lower side panels (Fig. 6-9). Simply hold the upper and lower side panels over the end panel and interior braces and flow liquid cement for plastics along the seam with a drafting pen or with an old No. 1-size paint brush. Also assemble the remaining side panel pieces to the back of the cab panel, but do *not* yet measure or cut any of the many sloping front panels.

Cut the panels for the boxed-in chassis using the plans and photographs (Figs. 6-6 through 6-13) to help you to determine the proper shapes and sizes. Assemble the chassis and glue the previously assembled side panel "box" and cab to the chassis (Fig. 6-10). The single sloping front panel that is shown in place in Fig. 6-10 can be cut oversize and glued in place to the ends of the cab side panels; when the glue is dry, the sides of that sloping panel can be trimmed flush with the corners of the sides.

The top of the hood is another panel that must be "developed," using the top view (Fig. 6-6) for its width and the corner locations and the side view (Fig. 6-6) for its length and for finalizing the corner locations. The angled

Fig. 6-8 The panels that angle away from vertical must be "developed" before they are cut from the plastic.

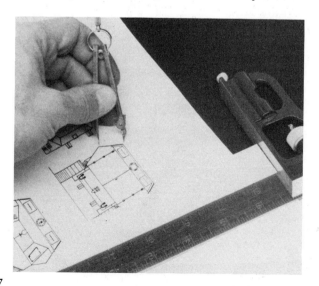

MODEL–BUILDING PROJECTS

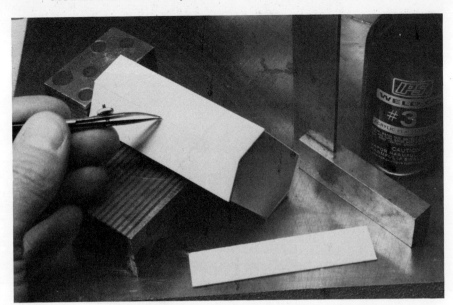

Fig. 6-9 Use an older style drafting pen or a paint brush to apply the liquid cement for plastics to the joints.

Fig. 6-10 The sloping front panel is in place on the cab and the interior seats and pedals are installed. Photo is by Dan Wilson of Miniature World, Colorado Springs.

Fig. 6-11 The top of the hood is next, followed by the panel at the front of the hood. Photo by Dan Wilson of Miniature World, Colorado Springs.

Fig. 6-12 The front wheels and tires and the track and suspension system are all from plastic kits. Photo is by Dan Wilson of Miniature World, Colorado Springs.

outer edges of the hood can then be drawn in place using the corners to determine where the lines should go. Score and break the hood from the sheet of plastic and glue it in place. Use this same technique to fabricate the small piece that forms the front of the hood and glue it in place (Fig. 6-11). The front bumper panel can be traced directly from the plans, so that part can be cut from the styrene sheet and glued in place.

The seat, steering wheel, foot pedals, and, perhaps, some radio gear can be "borrowed" from another kit. It's easier to install these parts before the windshield is in place. The windshield panel itself can be cut to the proper height, using the side view (Fig. 6-6) for the proper dimensions, and about 1/16 inch longer than necessary. Glue the piece in place and file the side angles to match the side panel after the glue is dry.

The suspension components from the "Chi-Ha" tank kit must be filed down in width so they will locate the kit's track properly. Carefully test-fit all of the pieces before you cement

Fig. 6-13 Support the hull on blocks until the cement that holds the track and front axle is completely dry. Photo by Dan Wilson of Miniature World, Colorado Springs.

Fig. 6-14 The completed Japanese "Ho-Ha" Type 1 half-track.

any of them in place. When installing the suspension and track, block up the completed body so it matches the side view in Fig. 6-6. The track and suspension can then be glued in place to support the body at the correct angle and height. Cut the front axle from scrap sprue and install it and the front wheels and tires at the same time. The bottom surfaces of the front tires should be filed flat where they contact the "road" to help give the impression that the vehicle really is packing the weight of armor plate.

Most half-tracks and other armored personnel carriers are a jumble of straps, boxes, cans, and tarps. An open-top vehicle, like this one, will also have brackets and straps to carry the rolled-up tarps that cover the top. One of those tarps and a spare parts or tool box are located on the rear panel; the second tarp is draped over some scraps of plastic inside the body. The tarps are pieces of facial tissue spray-painted a light beige or khaki color. The camouflage pattern on this particular vehicle can be simulated most effectively if it is hand painted to duplicate the rather crude color separations found on the actual vehicle.

Chapter 7
Trains

RAILROADS are romantic in ways you would never expect of a transportation system intended for nothing more than moving people or products from one place to another. There are about as many model railroaders in the world as there are model airplane enthusiasts and the hobby is growing at a faster rate than the real railroads are being abandoned.

Many model railroaders build their miniatures so they can simply display them on a shelf, but almost everyone in the hobby hopes to one day put static models into action. Few model railroaders, today, plan on having a purely static hobby; locomotives and rolling stock are meant to be moved as trains, not rested on a shelf. For the model railroader, then, the locomotives and cars are merely the stars of a universe that will one day include tracks, structures, and scenery.

Most of the models themselves are made from the same types of materials that are used for the majority of the rocket, aircraft, automobile, ship, and armor miniatures—plastic. Railroad modelers do tend to carry on some of the traditions of the old-time modelers in making at least some models from the same type of materials as the prototype. Given an infinite budget and unlimited spare time, most model railroaders prefer metal locomotives, wooden cars, and wooden structures for prototypes made of metal or wood. Both approaches to the hobby have their advantages, as you'll see here and in Chapter 8.

The Railroad Empire

Model railroading is one of the most versatile branches of modeling because the modeler's goal is to create an entire railroad scene including the trains, structures, scenery, and fine details such as people, fences, and real-life "clutter." This same approach has spread into other segments of the model-building hobby; builders of static model aircraft, automobiles, ships, and, particularly, armor vehicles now mount their prize models on plywood bases and surround them with the same type of scenery that model railroaders use.

The goal of a model railroader is an operating layout that at least fills a spare room if not an entire basement, attic, or garage. Modelers who are faced with temporary lack of space create model railroad dioramas built around a single structure or scene. If as little as 2' × 4' is available, a model railroader can create an operating portion of a model railroad, which is referred to as a "module." The modular model railroad can be incorporated into the future "dream" layout or connected to other modules to make a portable model railroad.

The concept of a constant scale is probably more important to a model railroader than to other modelers because the railroader wants to be able to blend every model he or she builds into a complete layout with scenery. With the exception of the buildings and scenery, the model railroader also expects all of his locomotives to operate and his rolling stock to be able to roll behind them. Generally speaking, model railroaders prefer relatively small models so that they can capture more of the "world" in what is never enough space. HO scale (1/87) is by far the most popular; about 80 percent of model railroaders prefer this size. N scale (1/160) is the next most popular size, with about 12 percent of active model railroaders building or collecting in this size. About 5 percent of model railroaders prefer O scale. There are also some small percentages of model railroaders building in S scale and others who build large-scale models for operation outdoors.

Materials for Assembling Railroad Models

The ready-to-run models represent a much larger proportion of the railroad hobby than in other modeling areas. The modeler who wants to build a complete operating railroad can buy his locomotives and rolling stock ready to run, use track that can be quickly nailed to cork roadbed, and decorate the system with quick-to-build plastic structure kits and ready-made or kit-built trees and shrubbery. Most of the hours are spent on the whole rather than on the parts.

The other extreme of model building is also well represented in the hobby; hundreds of modelers build their locomotives from sheets of brass with investment or "lost wax" detail castings equal to the quality of fine jewelry. Still others buy their locomotives but build wooden cars and structures stick by stick with scale lumber and precise castings for windows and trim details. There is also an ever-growing selection of building and freight car parts in plastic as well as plastic strips the size of scale lumber (from Evergreen) to enable the model railroader to use this modern material.

Model railroaders use the techniques described in Chapters 9, 10, and 11 for scratch-building plastic, wood, and metal models, respectively. There is no "standard" material for model railroad kits, although plastic is certainly the most popular. The plastic kit assembly tips in Chapter 1 apply directly to most model railroad kits. Many of the locomotives and freight cars have prepainted bodies, but some of these kits and almost all structure kits must be painted as described in Chapter 2. The weathering techniques in Chapter 6 are certainly appropriate if you want to match the usually dirty condition of most railroad equipment. Chapter 8 describes the kit-building techniques used in most of the model railroad structure kits.

Learning from Kits

There are basically three types of kits available for model railroad locomotives, cars, and structures: the simple kits that virtually snap together, slightly more complex kits that require glue and painting but virtually no fitting, and the "craftsman" kits that include some precut parts but all the details and trim must be cut to fit by the modeler. You can learn to build just about any type of railroad model from scratch by mastering the techniques of building these three types of kits.

Fig. 7-1 This 1/87 scale caboose is a combination of kit parts and scratch-built plastic sides and ends.

Some craftsman-style kits are especially helpful in teaching the assembly and fitting skills needed to scratchbuild your own models. Before you attempt to build any of the craftsman-style kits, though, you should complete several different types of the simple plastic kits and, perhaps, even build a few kit conversions to understand how to cement, cut, fit, assemble, and paint plastic or wood kits. There are no simple metal kits, but the experience you gain building with plastic or wood will be most helpful in training your fingers to work with tools and small parts. There are a tremendous number of wood (and cardboard) craftsman kits with metal or plastic detail castings.

The Silver Streak/Walthers freight car kits are a good place to start learning to build your models; when you're satisfied with the results of those efforts, move on to the more complex LaBelle Woodworking or Northeastern Scale Models kits. If you want to learn to scratchbuild with plastics, start with one of the E & B Valley HO scale kits and progress on to the Evergreen or Crossing Gate Models kits. The etched metal freight car kits offered by Quality Craft can be assembled with either cyanoacrylate cement or solder. Brass locomotive kits, however, require soldering and a considerable amount of time and patience. The Kemtron HOn3 and On3 narrow gauge 2-8-0 kits are some of the easiest of these complex craftsman-style kits. The Kemtron Wabash mogul 2-6-0 and a variety of limited-production brass kits from Precision Scale Co. are offered in both HO and O scale. These Kemtron and Precision Scale kits include precut and preshaped brass and castings, so all of the difficult work is done; there are thousands of solder joints to be made and each part must be fitted precisely.

A Caboose Kit Conversion

The Colorado Midland sidedoor caboose is an example of a wooden caboose style that was extremely popular with many railroads around the turn of the century. Even the Santa Fe Railway had an almost identical caboose. There is no kit to build a precise replica of this model, but the HO scale Model Die Casting No. 3400 all-plastic sidedoor caboose is close. You can learn some of the fundamental techniques of kit conversions for HO scale models with this project. Briefly, the conversion involves cutting the end platforms from the floor and the corresponding overhang from the roof with a razor saw. The kit's sides and ends are then filed to remove about .010 inch of the surface to make room for an overlay of new sides and ends cut from Evergreen's .020-inch thick scribed "car siding" (No. 2037) styrene sheet plastic. Use the score-and-break technique from Chapter 9 to cut the sides and ends from the plastic.

First, cut the ends, about 1/16 inch wider than the ends on the model; glue them in place and then file them to match the reduced width of the model. Cut the sides to match the height of the model but make them 1/32 inch longer and file their length flush with the ends after

Fig. 7-2 The Colorado Midland's 1890-era wooden caboose in 1/87 scale. Courtesy of L. P. Schrenk.

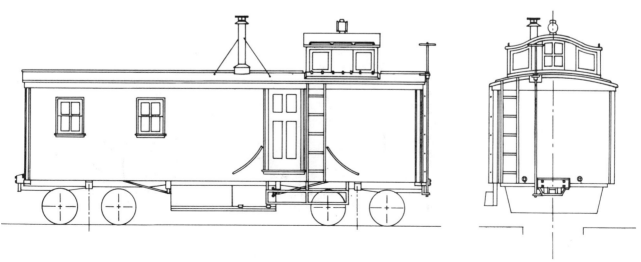

the sides are glued in place. The windows and doors must be cut into the sides *before* the sides are glued in place.

Cutting Windows and Doors

Fit the overall width and length of the sides to the MDC model. Use Larry Schrenk's plans for the Colorado Midland caboose (Fig. 7-2) to locate the exact outline of the windows and doors as described in Fig. 6-7. I suggest you use Grandt Line's No. 5059 windows and locate the door so you can use the molded-in panels and windows in the MDC kit—if you do, you may need to relocate some of the marks and lines for those door openings in the Evergreen sides. The windows in the MDC caboose are totally incorrect, so you'll have to file their openings oversize to provide an "open" area behind the windows in the Evergreen sides. The easiest way to cut the windows is to drill a ¼-inch hole in the center of each opening and then carve the opening to within 1/32 inch of the lines. Finish the opening with a flat jeweler's file to make a precise match for the Grandt 5059 windows (Fig. 7-3). If you are making a number of windows, you may want to use the punch method described in Chapter 9. Larger openings, like the doors, can be made by simply scoring and breaking the plastic. Make light pencil marks on the backs of the plastic to indicate where the top and bottom sills are located against the larger side panels before you remove them.

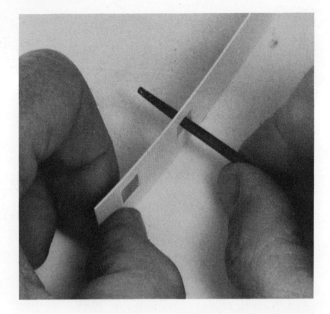

Fig. 7-3 Punch or gouge ¼-inch holes in the center of each window, then file the edges to shape with a rectangular jeweler's file.

Score and break the door panel along the vertical lines (Fig. 7-4), then snap off the upper and lower sills (Fig. 7-5). The sides can be assembled by cementing the upper and lower sills back into place with the pencil marks to help align them (Fig. 7-6).

Detailing Plastic Models

The complete range of plastic strips from Evergreen includes both inch-size strips such

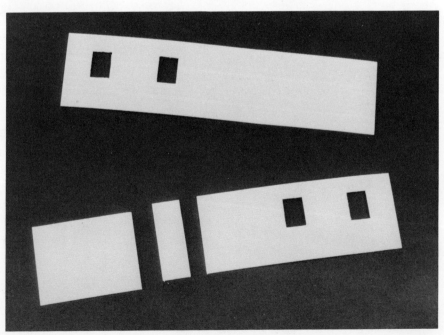

Fig. 7-4 Score and break the sides to remove the doors from the very top and bottom of each side.

Fig. 7-5 Use the score-and-break technique to remove the top and bottom door sills from the vertical door piece.

as .015" × .040" and precise HO and O scale strips such as scale-size 2" × 8" material. The smaller sizes of ABS plastic angles, channel, and round plastic strips from Plastruct and Model Parts are also most helpful in providing styrene-like ABS plastic details that can be quickly and easily cemented together.

Hair-size pieces of plastic can be made by holding a 6-inch length of leftover plastic sprue over a candle flame; as soon as the flame softens the plastic, pull or stretch it and you'll find it will stretch into almost any diameter strand of plastic you might need. The diameter of the stretched-sprue depends on how fast you pull it after the flame has softened it. It takes a bit of practice to know when the sprue is soft enough to stretch and how fast to pull it. In time, you'll be able to make stretched-sprue "wires" as small as .010 inch every time you try. The stretched sprues are best for details that are to be glued directly to the model like simulated electrical wires or pipes. For free-standing wirelike details, it's best to use the steel piano wire that is sold by model airplane shops. Drill holes in the plastic with a drill bit to match the diameter of the wire (plus .001 inch or so). The plastic is soft enough to be drilled by hand if you hold the drill bit in a pin vise (Fig. 7-7). X-Acto and General Tool make pin vises with chucks small enough to hold even the hair-size No. 80 drill bits. Make a light dent in the plastic with a compass point to keep the drill bit in position, then rotate the pin vise (with the drilled bit chucked tightly in place) to drill the hole. Hold the wire details in place with a drop of one of the "instant" cyanoacrylate cements such as Hot Stuff.

Most of the details on this Colorado Midland caboose were included in the MDC kit, but the ladders, roof walk, steps, and tool boxes could have been fabricated from Evergreen plastic strips if necessary. The cupola with the ess-curve is typical of pre-1900 cabooses of this type; it's included only in the No. 3444 MDC

Fig. 7-6 Cement the top and bottom sills and the sides over the thinned-down sides and ends of the MDC plastic caboose body.

Fig. 7-7 Use a pin vise to hold the drill bit while you drill holes in the plastic sides to match the diameter of the wire grab irons.

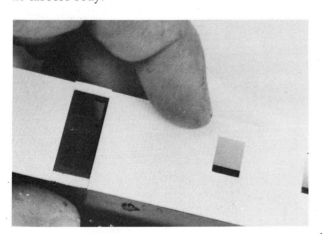

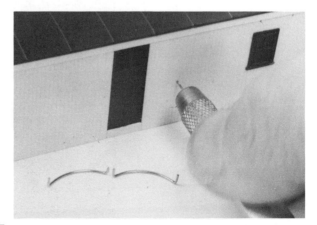

MODEL–BUILDING PROJECTS

Fig. 7-8 The MDC caboose kits can supply either a later style square cupola (*on the model*) or the earlier arched-roof style (*right*).

caboose kit. The black cupola on the caboose is a more modern style that the railroads fitted onto these cars after 1900; this is the only cupola available in the No. 3400 MDC kit. If you wish to build this conversion with the older style cupola, purchase an MDC No. 3444 kit with a pair of Grandt's No. 5063 caboose doors to use for side doors on the model.

Etched Brass Kits

The concept of using acid to etch brass has been used by the printing industry in making photo-engraved printing plates for decades. More recently, the concept has been applied to produce the printed circuit boards used in every modern electronic device and by hobbyists to build small radios and other electronic kits. Now, the principle of etching precisely shaped patterns in brass sheets is being applied to model railroading. The cabs and tenders of almost all the newest all-brass imported steam and diesel ready-to-run locomotives from the Orient are etched brass. Quality Craft sells several HO and O scale American-made caboose and freight car kits that are etched brass. The relative inexpensiveness of this method of cutting and detailing sheet metal makes it an excellent method for producing small batches of parts to meet the specialized demands of model railroaders. The etched brass process is explained in detail in Chapter 11 for those who might want to use this method to make their own parts for conversions or scratchbuilding projects.

Harold Mellor offers several conversion kits to fit the HO scale steam locomotives sold by Mantua and Tyco. His cab conversion kits include an etched brass panel to provide a one-

Fig. 7-9 Remove the superstructure from the Tyco/Mantua 2-6-0 to install the Mellor etched-brass cab (Fig. 7-10).

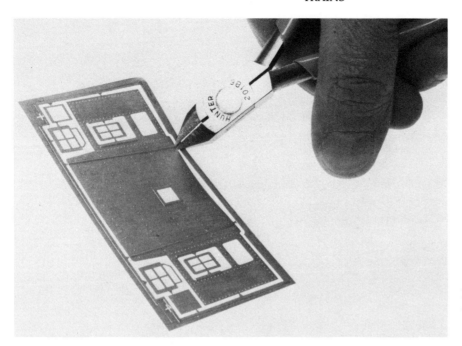

Fig. 7-10 Use flush-cut diagonal cutters to carefully cut the tabs that hold the cab pieces within the etched brass frame.

piece roof and sides for the Mantua/Tyco cab; the plastic cab provides the front and rear panels for the Mellor conversions. The etched brass sides are longer than the Mantua/Tyco cab, so the stock cab must be cut in half (crosswise) and the windows enlarged so they are not visible through the etched brass windows in the new sides. New etched brass side windows and a roof hatch are part of the kit, and strips of brass wire are included for the drip rail that often appears on the roof of such cabs and for the elbow rests on the window sills (Fig. 7-9).

To build the conversion cab, the etched brass parts are merely cut from the sheet with a pair of small diagonal cutters (Fig. 7-10). Diagonal cutters with "flush-cut" tips will leave only a trace of the attaching tabs on the parts. File smooth the edges of all the parts. Roll the roof of the cab very carefully, using

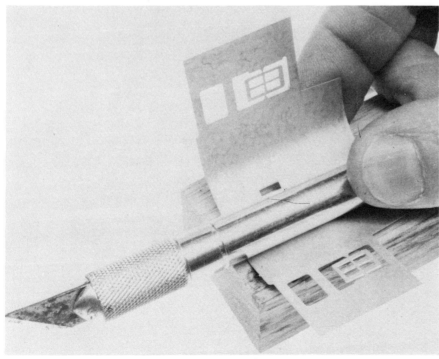

Fig. 7-11 Shape the curve in the cab roof over a round knife blade so the brass is not kinked or creased.

MODEL–BUILDING PROJECTS

Fig. 7-12 The new cab and Mellor's "Consolidation" kit's extra driver completely transform the appearance of the Tyco/Mantua locomotive. Photo courtesy of Harold Mellor.

the handle of a hobby knife to wrap the brass around (Fig. 7-11) so you don't accidentally kink the brass. Attach the brass roof and sides to the two halves of the Mantua/Tyco cab with either 5-minute epoxy or Goldberg's Super Jet. Use the same cement to attach the brass cab windows, elbow rests, roof vent, and drip rails. The cab itself is installed on the boiler using the attaching screw and hole that held the original cab in place.

The completed locomotive (Fig. 7-12) is a Mantua 2-6-2 that has been fitted with the

Fig. 7-13 The prototype for our 1/87 scale model, the Santa Fe Railway's 406 class 4-6-0 in about 1890. Courtesy of Denver Public Library.

TRAINS

Fig. 7-14 1/87-scale plans for those who want to photoetch a brass cab. Courtesy of Robert Sloan.

Mellor "Consolidation" kit to provide a fourth pair of drivers in place of the two-wheel trailing truck. This model has also been detailed with Cal Scale and Kemtron lost wax brass casting details including a new headlight, bell, whistle, and steam generator, as well as some additional wires to simulate pipes.

A Cab Conversion

You can use the drawings in Fig. 7-14 to make your own etched-brass cab. If you'd rather purchase a ready-made set of etchings, they are available in HO or S scale from Robert Sloan.

This cab uses a much simpler type of etching than the Mellor parts because all of the lines are etched right through the brass sheet. The Mellor etchings include surface rivet details that are only etched part-way through the brass to leave the rivets standing clear of the surface.

The etched brass cab in Fig. 7-14 utilizes two layers of brass for the sides and ends to obtain the recessed effect of the window frames and door frames. The cab is a precise replica of a Standard Baldwin Locomotive Works all-wood design that was used on hundreds of locomotives before the turn of the century. It was copied from the cab on a Colorado Midland locomotive, but it is typical of the cabs used by almost every railroad in America. The cab makes it possible to convert the Mantua or Tyco 4-6-0 (or the almost identical 4-8-0) into an almost exact-scale replica of the Santa Fe Railway's 406 class locomotives (Fig. 7-13) with the addition of a new set of domes and some

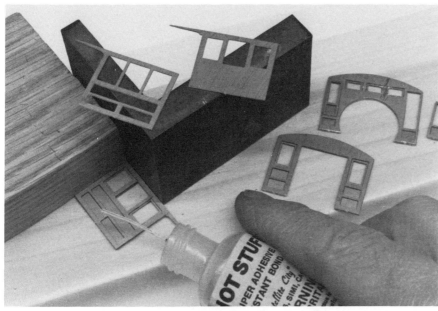

Fig. 7-15 The Sloan cab etchings use two layers to simulate the wood paneling. Block the two in a corner while you apply cement.

MODEL–BUILDING PROJECTS

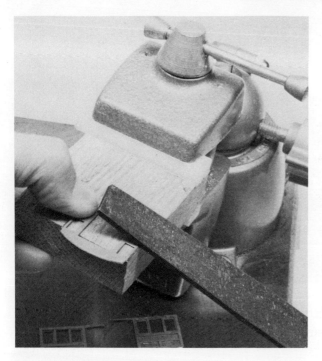

Fig. 7-16 Each of the corners must be filed to a 45 degree angle before the sides and ends are assembled together.

wire details and decals. Use the plans in Fig. 7-14 to produce your own etchings as described in Chapter 12 or purchase Sloan's.

Cut the parts from the etching's scrap brass and file the edges of the parts smooth. Assemble the inner and outer pieces for the two sides, the front, and the rear of the cab, but be sure to make a right and left side (the parts are reversible). You can assemble the cab with either solder or cyanoacrylate cement. I used conventional cyanoacrylate (Hot Stuff; see Fig. 7-15) for laminating the front, back, and sides and the thicker Super Jet to assemble the parts (Fig. 7-17). You *must* file the inner corners to a 45 degree bevel angle (Fig. 7-16) with this particular set of etchings. Some etchings allow the sides and front to overlap, however, so read any instructions furnished with the kits you purchase.

No roof is included in the set of Sloan etchings; you must cut one from a .016-inch-thick brass sheet .020 inch larger than the upper edges of the curved cab ends. The K & S brand brass sheet stock is available at most hobby shops or you could substitute .015-inch-thick Evergreen sheet plastic.

The Mantua/Tyco boiler must be filed and trimmed a bit to fit the Sloan cab in place of the model's stock cab (Fig. 7-18). You can turn new domes and a new stack as described in Chapter 11 or purchase these parts from firms such as Arbour or Kemtron. Mount the cab on the boiler with 5-minute epoxy or Super Jet, paint the model and apply new decals (Fig. 7-19).

Building a Steam Locomotive

It takes a considerable amount of skill to build any scale model from raw materials, but few models are as complex as replicas of steam locomotives. You'll need to develop your metal-working skills, particularly the ability to shape sheet metal and to solder several parts together without loosening the parts that were previously soldered during the process. It is possible to build a steam locomotive from plastic, especially if you can remove some of the domes and other details you need from one of the injection-molded models. There's absolutely no reason that the thousands of brass detail parts cannot be used on a plastic locomotive. Five-minute epoxy or Super Jet will hold the metal details to the plastic almost as well as solder—larger details can be attached with small screws or pins from inside the model. The rivet-embossing technique described for use with plastic in Chapter 9 will work equally well with metal.

The most important "secret" that any experienced scratchbuilder can provide is the art of scrounging; you can save yourself hundreds of hours of time if, for example, you can find a smokebox front that is correct for your proposed steam locomotive or a piece of brass or

Fig. 7-17 A few steel or wood blocks can be used to ensure square corners. Protect the blocks from the cement with wax paper.

Fig. 7-18 The Tyco/Mantua boiler must be filed slightly to clear the hole in the new cab. The original cab is on the left.

plastic tubing that is the correct diameter for the boiler. Innovation is also important; if that tubing is not the right size, then you may be able to wrap one or more layers of sheet plastic (or sheet brass) around it to build up the diameter.

Researching Model Locomotives

Today, there is seldom any advantage in making a detail part (like that smokebox front) if you can find one already made. Before you begin to scratchbuild any project, find out what kits might have appropriate parts and determine how many detail castings are available to match those on the prototype. You won't know what parts you're looking for, of course, until you are thoroughly familiar with the prototype for your model.

When you're searching through books and magazines for photographs or plans, try to find out if any examples may still survive. Few model railroaders know, for instance, that there is a surviving example of the most popular HO scale imported brass steam locomotive in history; one of the Santa Fe Railway's 1950-class 2-8-0 consolidations rests in a park in Pauls Valley, Oklahoma. The prototypes for Albert Hetzel's 3800-class Santa Fe Railway 2-10-2 locomotives were all scrapped; he used over 50 different photographs of the prototype

Fig. 7-19 The new cab, dome, stack, and black paint completely alter the appearance of the Tyco/Mantua 4-6-0 to match the Santa Fe engine.

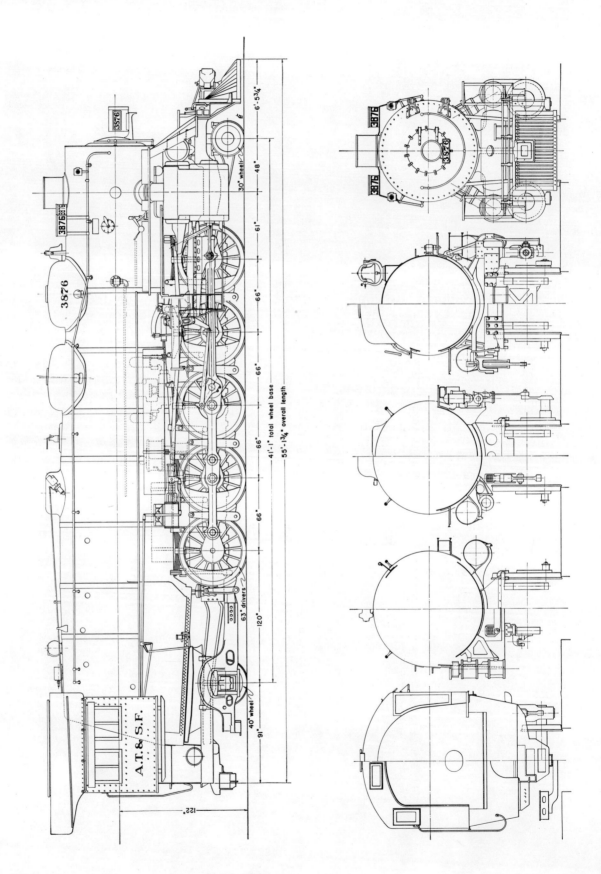

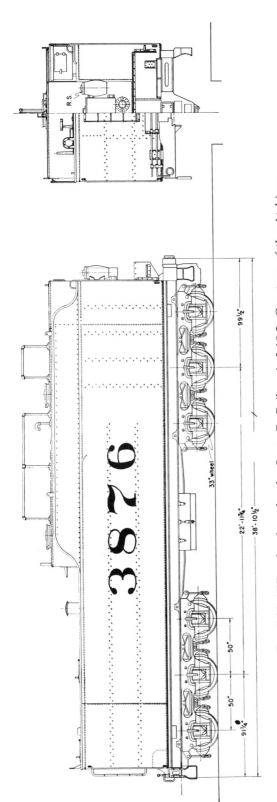

Fig. 7-20 1/87-scale plans for the Santa Fe Railway's 2-10-2. Courtesy of the Atchison, Topeka and Santa Fe Railway Co.

MODEL–BUILDING PROJECTS

Fig. 7-21 The boiler, cab, and domes on Albert Hetzel's 1/48 scale Santa Fe 2-10-2 are soldered together and ready for the fine details.

Fig. 7-22 The smaller details are commercial lost-wax brass castings, but the cab, boiler, domes, frame, and tender are all handmade.

Fig. 7-23 The pipe joints, wrapped pipe, plugs, clamps, and brackets are all available from firms that sell detail castings for trains.

Fig. 7-24 Albert Hetzel used a male and female punch to emboss each of the rivets into the .015-inch-thick brass tender shell.

as well as these plans (Fig. 7-20, enlarged to 1/48 scale) to build his O scale model.

A Handmade Brass Model

Albert Hetzel's O scale Santa Fe Railway 2-10-2 is an example of what most of us would call the "ultimate" in handmade or scratch-built miniatures. Fifty years ago, this model would have been made from nothing more than blocks and sheets of brass with, perhaps, some wheel or driver centers cast in sand molds. Today, there are firms like Precision Scale Company, the Back Shop, Grandt Line, American Locomotive Works, and Kemtron that offer detail castings in HO and O scales and like Precision Investment Associates and Cal Scale that offer HO scale detail parts as incredibly detailed lost wax ("investment") brass castings. It would take a skilled craftsman a lifetime to come even close to the details in the castings on Hetzel's locomotive. The cumulative number of hours spent by the master sculptors who carved each of the patterns for the parts on this one locomotive actually totals about 80 years!!

During his 40 years in the hobby Hetzel also developed the ability to scrounge; he managed to collect the drivers, trailing truck, stack, and a few detail castings for his model when some similar 2-8-2 all-brass locomotives were manufactured in Japan and Korea a few years ago. In some instances, the importers of such locomotives will import only a few such spare parts that are available only at the time the complete locomotive is imported.

Hetzel's ingenuity came into play in duplicating the tender trucks, smokebox front headlight, and bracket and water filler hatch on the tender; he made molds from RTV rubber for those parts and cast duplicates in aluminum-filled Duro brand epoxy as described in Chapter 12. The molds were made using the parts from one of those same 2-8-2 imports. Hertzel did saw and file the frames by hand, but he used the 2-8-2 frame as a pattern, adding another driver and adjusting the dimensions of the new frame to match those of the prototype.

Hetzel's 2-10-2 locomotive qualifies as a scratch-built locomotive, however, because he made the major parts, including the frames, cylinders, boiler, cab, domes, and tender, from sheet and bar brass. Many modelers dream of building such a model steam locomotive, ship, tank, or automobile. I would love to explain to you, step-by-step, just how each part of Hetzel's masterpiece was created; but that is not how he learned to build brass models and it's certainly not the way for you to learn.

The *only* way to learn to build in brass is the way Hetzel learned—by actually doing it. He built three of Kemtron's On3 C-16 all-brass kits and two more of the O scale locomotive kits from Precision Scale Company before he attempted to tackle this one. If you have assembled a couple of these kits, you should find any "missing" information on scratchbuilding in brass in these pages. Chapter 11 will certainly be the most helpful, but there is also important information in every other chapter.

Hetzel's 2-10-2 locomotive has a hand-cut cab and tender and firebox with rivets embossed exactly as described for working with plastic sheet stock in Chapter 11 except, of course, that these parts are made from brass sheet stock on this model. The only reason Hetzel built the locomotive was that he could not buy it ready-built or as a kit; even though he did buy every part possible it still took over a year of weeknights and weekends to complete the model!

That's the sensible approach to building from scratch. In fact, most professional model builders would have been even less romantic and more concerned with the reality of such a project; 9 out of 10 of the professional modelers I have talked to would have used a commercial chassis for such a model and built the superstructure and tender from plastic rather than brass. Even Albert Hetzel might have taken this approach if he had not already collected most of the chassis and detail parts in metal.

The moral, here, is simple: don't start a project just to "make work" for yourself. The modern world of model building offers an incredible array of time-saving detail parts and labor-saving materials. The standard of excellence in today's miniatures is also much higher than it used to be as a result of such parts and details. The steam locomotive model fabricated from stovepipe, thread spools, and broom handles was impressive, indeed, in 1910, but today such a model is merely crude when considered as anything but an antique.

Chapter 8
Buildings

STRUCTURES might seem like a rather mundane subject for a miniature, especially when compared with the implied action of a tank or sailing ship or the realistic movements of a miniature electric train or radio-controlled race car or airplane. The structure as a subject for a model is very closely related to the still life subjects that an artist might select for an oil painting. The structure offers some unique opportunities to the skilled modeler to suggest that the building is (or once was) inhabited or worked in by living people. The exact scale reproduction of the bricks, shingles, windows, and doors is only the starting point for a realistic model of any building. The small details like rain gutters and downspouts, doorknobs, electrical meters and wires, the surrounding landscape, and just a touch of weathering are all as important as the basic model.

The overall shape of the structure is important as well, but it is not necessary to construct the building to precisely the overall *size* of the prototype. The study of "selective compression" of the prototype to make a model could fill this chapter; briefly, the sizes of all doors and windows are kept to exact scale but the number of doors and windows can be reduced to allow the model to be a bit shorter, narrower, or even lower. A five-stall railroad roundhouse model, for example, will look virtually as realistic as an eight-stall replica of a prototype without losing any of the prototype's flavor. The goal of anyone modeling a building should be to capture the character and atmosphere of the real thing. The exceptions to such a rule, of course, are the engineering and architectural models that are themselves "prototypes" for the full-size buildings and industries they represent.

Most of the structures that are assembled by hobbyists are intended to serve as scenery for model railroads. There are hundreds of plastic, wood, and cardboard kits available in HO (1/87), N(1/160), and 0(1/48) scale that include the stations, railroad buildings, and industries you'd expect to see on a railroad as well as some unlikely models such as cathedrals, castles, log cabins, cottages, Victorian homes, and a wide range of downtown businesses and houses. There is an entire hobby centered on 1/12 scale buildings that includes a variety of houses and stores as well as a range of windows and doors and other details that actually operate—the hobby, of course, is dollhouse miniatures but it is now a scale model rather than a toy category. There are also several plastic and cutout paper models in 1/700 and 1/1200 scale to complement dockside scenes for waterline ship model dioramas. Many of the 1/72, 1/35, and 1/48 scale plastic "accessory" packs for armored fighting vehicle dioramas include details (like telephone poles and tin pails) that can be used with HO or O scale structures for dioramas or in model railroad scenes.

Cardboard Models

If you became an active model builder before the 1970s, there's a good chance you discovered artists' Strathmore brand cardboard. This

Fig. 8-1 Leila Johnson used Houseworks brand windows to detail this Our House brand 1/12 scale dollhouse kit.

material is available in a variety of thicknesses or plies (layers) ranging from one-ply (.005 inch) to five-ply (.025 inch) in sheets 23" × 29" or 30" × 40". The material is very similar to the cheap cardboard used for packing shirts or for backing writing pads, but it is much denser. Strathmore will not fuzz along cuts like cheap cardboard. It does have a grain that you can feel by trying to flex the sheet along its long and short lengths; the grain runs parallel to the sides that are the stiffest. If you want to make any type of curved surface with the Strathmore, the grain must be at a right angle to the curve so the board will bend and not break.

I cannot recommend even Strathmore cardboard over sheet plastic for a model builder. The only time you might want to use cardboard rather than sheet plastic would be when high-intensity light that throws off extreme heat (or direct sunlight) will be used to display the models. The models made for use in television and motion pictures, for example, are often made of cardboard because the lighting would warp most plastics. For most modelers, however, sheet plastic can be used anywhere you would use Strathmore, and sheet plastic is easier to cut, quicker to glue, and less susceptible to warping than any cardboard. The Suncoast and some Suydam HO and O scale model railroad building kits are made of cardboard (with wood and metal details) so you might want to try building one or two kits made by these firms to see if you enjoy working with cardboard. Try one of the Evergreen brand plastic structure kits, though, before you decide that cardboard is best, even if you did learn to build models over a decade ago.

There are a growing number of models made of thin cardboard with the parts printed (usually in full color) directly on the cardboard. The cardboard used is really like the fine-quality one-ply Strathmore, but these models are called "paper" rather than cardboard. The construction techniques are virtually identical to those outlined in this chapter to build the cutout railroad station. The models must be carefully trimmed from the

Fig. 8-2 The two-dimensional surfaces of the windows and doors on these Dover Publishing cutout structures are not apparent in a photograph.

printed sheets with a hobby knife or single-edge razor blade. Some corners are merely folded over while others are formed by tabs that are to be cemented to the inside surfaces of the model.

Dover Publishing has a series of Western, New England, and Victorian buildings that are close to HO scale; 101 Productions offers several Victorian buildings, mostly in O scale, that are not precolored; and John Hathaway imports dozens of both precolored and simple line art cutout kits that include castles, historical landmarks, and a range of aircraft and ship models that are incredibly complex. D. M. Emmons also has some complex paper aircraft models. Cutout paper models are a hobby in their own right and the models can last a lifetime if you protect them, inside and out, with a light spray-on coat of Testors' Dullcote or Glosscote.

Any of these models can also serve as excellent plans (and patterns) for scratch-built miniatures in sheet plastic, wood, or even Strathmore. In some instances, the cutout material can serve as the walls for the model with new three-dimensional cast metal or plastic windows and doors.

Cardboard Kits

The only kits available to the modeler in the decades from 1930 to 1960 were made from precolored cardboard about .030 inches thick with die-cut windows. The doors were generally printed on the surface of the cardboard or on separate sheets of thinner material. Old timers will recall the Ideal and Skyline brands that have been out of production for 20 years. The windows for these kits were simply celluloid or other clear plastic with the mullions printed on them in ink.

Today, similar HO scale kits are made in England by Super Quick and imported by John Hathaway. The Builder Plus series of N and HO building kits are similar in construction, but they have a thinner wall material (about .010 inch) and some, apparently made with a

Fig. 8-3 John Hathaway imports dozens of cutout, precolored structure models, including this "Medieval Castle" kit that was assembled by Mike Czibovic.

full-color photographic process, have a truly realistic brick appearance. Polks imports the Builder Plus series into America. Either of these brands is well worth detailing with cast metal or plastic windows and doors.

Building a Cutout Station

This HO scale railroad freight and passenger station will give you some idea of how a paper model is assembled. I suggest that you pho-

Fig. 8-4 Use a steel ruler to guide your knife blade when cutting out paper models. Work on a hard surface for clean cuts.

BUILDINGS

Fig. 8-5 Two layers of three-ply (.015-inch-thick) Strathmore were used to make the walls in Jim Malisch's model store.

tocopy the pages and glue the paper copies to a sheet of single-ply Strathmore board with either rubber cement or, better, with 3M brand Sprayment adhesive. This step is not necessary (or recommended) with the cutout kits because the cardboard is already thick enough.

Use a hobby knife to cut through the paper and Strathmore with a steel ruler to guide every cut so it is perfectly straight. Scissors have a tendency to rip the material even if they are sharpened properly. If you are cutting thicker cardboard (two, three, four, or five-ply

Fig. 8-6 Betty Dick used one-ply Strathmore for the windows and doors and three-ply for the walls of these stores.

Strathmore) it's best to use a single-edge razor blade or a surgical knife; the blades in hobby knives are so thick they will curl the thicker cardboard along the edges of each cut. Bend each of the corners and each of the tabs over a steel straightedge placed on the edge of the table.

These models can be assembled with white glue or with the plastic resin-based glues if you can devise a way to clamp the parts together until the glue dries. The cyanoacrylate cements may soak into the paper to create a spot, but you can use them if before assembling the parts of the model you spray them with Testors' Dullcote or Glosscote to seal the surface of the paper. Some of the paper models are precoated so this step is not always necessary. I've found Goodyear's Pliobond cement to be quite suitable, but you must be careful to keep the brown fluid from touching any of the visible portions of the model. Rubber cement is not strong enough for a permanent bond on many of the parts. If the model is not precolored, you can color it with a felt-tipped pen after it is assembled. You may smudge the colors if you color the parts before they are assembled. The felt-tipped colors are just translucent enough so the inked lines will show clearly; paint would hide the detail lines.

When the model is complete and colored, spray on one or two very light coats of Testors' Dullcote or Glosscote to seal the surfaces so moisture will not make the surfaces sag. Any of the paper or cardboard models will be much more resistant to warping if you reinforce all the walls along every vertical and horizontal corner with 1/8-inch square strips of basswood. The reinforcing strips can be added to the inside of the model just before the roof is installed. The strips can be held in place with the same type of glue you used to build the model.

Building with Wood

Wood is still the most popular material for model structure kits in 1/160, 1/87, 1/48, and even 1/12 scale. The wood kits generally fall into two types: those that use large sheets of

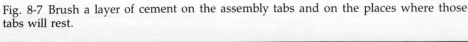

Fig. 8-7 Brush a layer of cement on the assembly tabs and on the places where those tabs will rest.

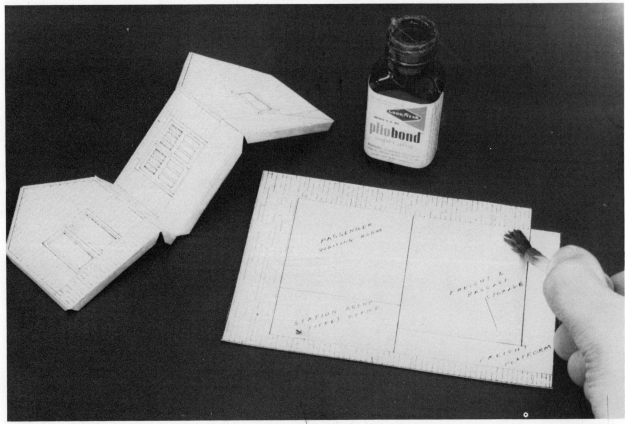

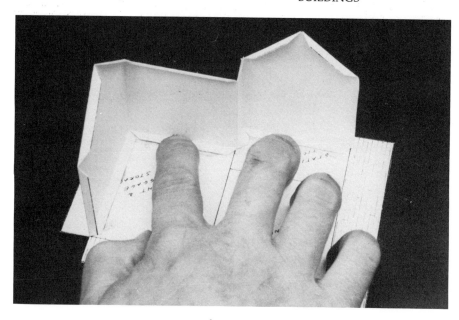

Fig. 8-8 Press the tabs tightly against the adjoining surfaces and hold them there by hand or with weights until the glue dries.

milled basswood for the walls and those that provide scale-size individual interior bracing (wall studs and ceiling rafters) and exterior surfaces as precut wood strips. Either type of kit makes the perfect "hands-on" learning tool for those who might want to build their own structures from scratch. Most of the kits feature die-cut walls with the windows and door openings punched and ready to be cut out with a quick push of a hobby knife in each corner. Some kits even feature cut-to-length wood strips for corner trim and details. Most of the modern kits include cast metal or plastic windows (in 1/160, 1/87, or 1/48 scale kits) or precut wood window and door moldings for working windows or doors (in 1/12 scale and some 1/48 scale kits).

The difference between a kit of this type and a scratch-built model is that the modeler does not have to do any difficult cutting and the model is completely engineered so that it really will make what it is supposed to. The kit maker has also collected all the parts for the modeler. Once you have learned how to build a kit of this type, however, you can duplicate the construction methods using the

Fig. 8-9 A few Preiser brand HO scale figures add a touch of life to this cutout railroad station.

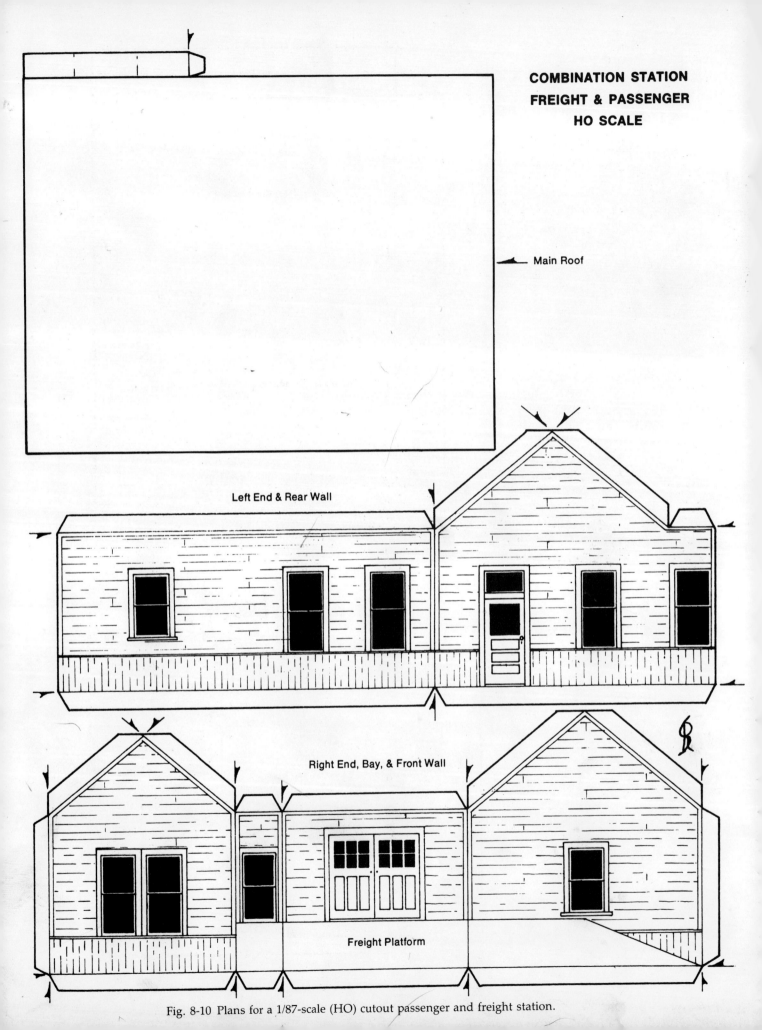

Fig. 8-10 Plans for a 1/87-scale (HO) cutout passenger and freight station.

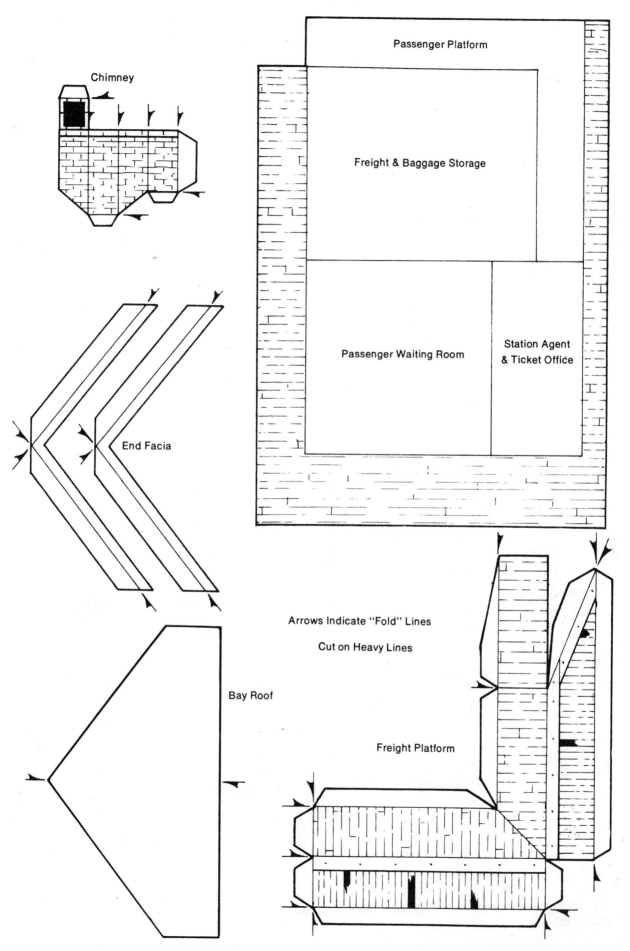

Fig. 8-11 Full-size plans for the floor and details of the 1/87 scale cutout passenger and freight station.

same materials that are in most of the kits. Northeastern milled-wood sheets for clapboard or board-and-batten style walls are stocked by many model railroad shops along with a selection of scale-size "lumber" in 12- or 24-inch strips. Basswood is the best material for structures; balsa wood is too soft and the grain is too pronounced. These stores also carry the same metal and plastic windows and doors that are included in most kits so you can collect your own "kit" of parts; all you need are a set of scale plans (see the back of this book) and the energy to cut all the windows, doors, and stripwood.

Simulating Brick and Stone

Brick and stone are the most difficult surfaces to simulate on a model structure. The historically proven method of simulating these materials is to use sheets of "brick paper" or "stone paper"—precolored paper with the lines and shadows printed on its surface. Brick paper looks fine in the two-dimensional world of a photograph, but it's simply not convincing in reality. The only brick papers that are acceptable are those made by Builder Plus in England and imported by Polks.

Even with these photographically reproduced bricks, however, you must trim individual bricks or rows of bricks from the paper and cement them to identical-size .010-inch thick strips of Strathmore to make window sills, lintels, and cornices. You are then faced with the problem of trying to match the color of the printed bricks on the edges of that .010-inch cardboard. The best way around the problem is to duplicate a prototype that really does have a solid panel of brick for all the walls with either concrete or stone lintels, sills, and cornices. Basswood can be carved to simulate stone or concrete and sealed with two coats of sanding sealer (sanding the surface between each coat with fine-grit sandpaper) to hide the grain. You may even use plastic trim parts from a kit for such details.

The best way to duplicate the three-dimensional appearance of brick or stone is to use three-dimensional material. Campbell offers a vacuum-formed stone material in plastic, and Holgate and Reynolds has a variety of brick, stone, concrete block, and shingle textures in a similar material. These plastic sheets must also be cut into individual strips and backed with .010-inch thick cardboard or styrene plastic strips for lintels, sills, and cornices.

There are, however, two ways around the problems if you use plastic sheets of brick or stone; you can avoid the problem by simulating stone or concrete sills, lintels, and cornices or you can "cheat" and use the plastic brick or stone sills, lintels, and cornices from one of the Vollmer, LytLer & LytLer, or ConCor/Heljan kits that has a matching brick or stone texture with separate details. Some of these kits have such details molded into the plastic wall

Fig. 8-12 Albert Hetzel used windows and walls from several Vollmer kits to make this large three-story warehouse in 1/87 scale.

panels, but some offer separate details. It would be impossible to match the color of any brick or stone "paper" with three-dimensional plastic bricks but, here, both the wall panels and the details will be three-dimensional and both can also be painted at the same time to blend perfectly together.

The three-dimensional bricks or stones are relatively simple to paint; cover the entire surface with a single coat of concrete or grout-color gray paint. When the paint dries, rub the brick or stone color over the surface with a rag so the color hits only the surfaces of the brick, leaving the concrete gray color where it should be, between the courses of brick or stone. The same technique can be used to paint or highlight the plastic roof shingle material.

Plastic Kit Conversions

The Evergreen brand styrene plastic strips provide one of the quickest methods of building a simulated-wood structure. The techniques for working with plastic in Chapter 9 should be used with Evergreen's material. The best way to develop your skills so you can scratchbuild in plastic is to assemble one of the Evergreen brand structure kits. Their HO scale structures are all plastic and use the same method of assembly that is featured in most of the wooden structure kits.

The difference between a kit of milled wood sides and strips and one of plastic is that the plastic kit can be assembled almost as fast as you can pick up the pieces. The liquid cement for plastics requires about 24 hours to dry com-

Fig. 8-14 Use slower drying Testors liquid cement for plastics when a number of strips (like these rafter ends) must be installed.

pletely but the joints are firm enough to allow you to go on to the next stage of the assembly in a matter of minutes. The white glue or plastic resin-based cement (like Titebond) used on wood models requires several hours to dry enough to allow the model to be moved. The square edges of the plastic strips are closer to scale, to simulate wood, than wood itself. The coarse grain of aged wood can be brushed onto the surface of the plastic pieces with a stiff wire brush before they are cemented together.

The Evergreen kits, like most craftsman kits, include wall, floor, and roof panels that are cut to size with all of the window and door openings precut. The small strips that are used for trim on all the corners, for exposed rafter ends, and around some of the doors can be cut to fit directly on the model (Fig. 8-13) or on the full-size plans furnished with the kit. The plastic is easier to work with than wood; again, simply score the strips with the knife and bend them downward to snap off the strip of plastic along the score line.

I suggest that you use two different brands of liquid cement for plastics to assemble a model made from an Evergreen kit or from any styrene sheet plastic. Testors liquid cement dries a bit more slowly than the cements from MicroScale, Plastruct, or Weld-On. Use the Testors where there is a very long seam or where a number of parts (like the exposed

Fig. 8-13 The thin plastic trim strips can be cut to length using just light pressure to score the plastic; the strip will break along that scored line.

MODEL-BUILDING PROJECTS

Fig. 8-15 Attach the windows and details to a scrap of wood *(bottom)* so they can be spray painted before they are installed.

Fig. 8-16 The parts from the Evergreen brand "Iron Works" were modified and rearranged to make this one-off industrial "foundry" structure.

rafter in Fig. 8-14) must be positioned in one area. The other brands work best for most other joints because they dry quickly. It is seldom necessary to apply the liquid cement before assembling the joint; merely hold the parts together and touch a drop of the liquid to the seam with a paint brush or a drafting pen and capillary action will draw the cement into and along the seam. Longer seams may require two or more applications to provide enough liquid to completely cover the seam. The parts can still be moved for a few minutes after the cement is in place so you can adjust the angle or fit to perfection.

Painting Tips

Most structures are fitted with windows, doors, and trim in a contrasting color to the walls and roof. It's possible to paint the trim after it is glued in place if you use a No. 000-size brush. However, there's an easier way—paint the windows, doors, and the plastic strips that will be used as trim *before* gluing them to the walls, and paint the walls at the same time you paint the trim. Attach the trim and small parts to a scrap of wood by folding masking tape back over itself so the sticky side is up. You can then paint the trim and the assembled walls of the structure with an airbrush or with aerosol cans (Fig. 8-15). The plastic cement works by dissolving bare plastic, so it will *not* work after the parts are painted. The painted windows, doors, and trim can be installed using cyanoacrylate cement (such as Super Jet). The clear plastic window "glass" can be installed in each window, using white glue, before the windows are installed in the walls.

The multipiece construction of the Evergreen kits makes it particularly easy to rearrange the walls and the various lean-tos to produce a model that is barely recognizable as being based on the kit. The foundry in Fig. 8-16 is a revised version of Evergreen's No. 804 Iron Works kit with the side walls shortened 2 scale feet to allow an overhanging roof on both the front and rear of the main building. The false front wall was then modified from a rectangle to a peaked-end to match the rear wall. The simulated tarpaper roof was made from strips of construction paper supplied in the kit. The streaks and weathering were applied with an airbrush to complete the model.

PART II
Building Materials and Techniques

Chapter 9
Plastics

THE most modern material made for model building is also one of the most maligned. Many modelers find it impossible to differentiate between the one-piece plastic toys sold in the dime stores and the $60 cathedral kits in precise HO scale. Plastic was once associated with "cheap" and that's a difficult image for some to forget. In truth, plastic is the best material for just about any model building project. Those who make their living building models cannot do without it. I've talked to over a hundred such professionals, and more than 90 of them prefer plastics when they have a choice.

The plastics used for model work are very stable over a tremendous range of humidity and temperature, which makes the material essentially unchangeable under the conditions in which most models are displayed. The styrene and ABS plastics used for kits and for scratchbuilders' supplies will warp and deteriorate in direct sunlight, at over 120°F, or when exposed to the heat of the lights used for television and motion picture photography. Few of us need worry about such extremes. For the modeler, plastic is virtually as strong as metal, it won't warp or sag, and it can be worked much easier and quicker than the softest wood or cardboard. Plastic can also be cut, drilled, machined, and embossed so it's perfect for simulating metal. It can also be distressed to simulate wood, brick, or stone, and it can be heat-formed in bends or into bubbles or vacuum-formed into complex shapes and textures. Best of all, the strips, plain sheets, milled sheets, and structural shapes in plastic are about the same price as similar wood or metal for model building purposes. It's not the perfect material for everything, but you'd be wise to heed the professionals' advice to consider plastic as your first choice.

Precision Work

Plastic holds an edge without splintering like wood or fuzzing like cardboard, so it is possible to drill holes as small as a No. 80 (.0135-inch diameter) and to work with strips equally small in either round (heat-stretched sprue) or rectangular (Evergreen brand strips) shapes. Industrial plastics supply firms can sometimes supply large blocks of plastic that can be carved and machined almost as accurately as brass but with just a fraction of the effort. The largest Evergreen strips, .125" × .250" (⅛" × ¼") can be laminated into blocks large enough for most model work using liquid cement for plastics. The ABS and styrene plastic most suitable for model work can be shaped in a small milling machine such as the Machinex 5, Unimat, or Sherline.

If you do not need the advantage of solvent-action cement joints, there are other plastics, such as Dupont's Lucite, that are even more adaptable to turning in lathes or cutting in milling machines. A large number of the patterns used for the injection-molded plastic kits are made of plastics like this in the pattern

Fig. 9-1 Precision measuring and alignment tools (*left to right*): combination square, machinist's square, surface plate, steel scale ruler, two rectangular steel blocks, PanaVise, PFM vernier calipers (HO scale) and 2-inch micrometer.

shops of the model kit manufacturers. The ready availability of milled and plain sheet styrene and ABS plastic, the lumber-size Evergreen styrene strips, and the Plastruct and Model Parts ABS plastic angles, channels, I-beams, and other structural shapes and tubes has been instrumental in upgrading the standards of scratch-built models during the past decade. Plastic cuts so easily that it is possible to maintain standards of fit and precision that were once only possible in metal with expensive machine tools.

There is no way to obtain the precision needed for a scale model using a yardstick as your only measuring tool. When applied to models, the term *precision* should probably be revised to *consistency* and alignment. It's not important that the two sides of a boxcar be accurate to a thousandth of an inch, but it is important for them to be *equal to one another* within a thousandth of an inch. When they are assembled, they must also be fitted squarely with the ends.

You need two types of tools to build accurate scale models from plastic (or any other material): tools that can measure lengths or diameters precisely and tools that help to align assembled parts so they are perfectly square. If you are a model railroader, I recommend the purchase of the PFM brand vernier calipers for either HO(1/87) or O(1/48) scale. The calipers are expensive but they are accurate to about ½ of a *scale inch!* You'll need a steel ruler of some kind, so you might as well buy one that is marked in scale feet and inches (HO, S, and O) like the 12-inch long General Tool product. If you cannot locate, or cannot afford, the PFM calipers, then purchase an inexpensive 2-inch micrometer such as the Sears' Craftsman brand. Any scale dimension can be converted to a decimal fraction of an inch so you can read it on a micrometer. The vernier calipers, micrometer, and scale ruler are the instruments that will help you to make all those important duplicate measurements exactly the same *every* time.

Square corners begin at the base of the model so you really should have a surface plate like the ground-smooth (No. 310) surface plate that doubles as a base for the PanaVise brand swivel-mount vise. You can substitute a ½-inch-thick piece of plate glass, but there is always the danger of chipping the glass. You may also be able to purchase a used surface plate from a machine tool supply firm. A small all-steel machinists' square, like the 4-inch No. 2704 sold by General, is extremely helpful but you can get by with the somewhat more awkward 12-inch carpenters' combination square. A pair of ground steel blocks about 1" × 2" × 2" can be extremely helpful in aligning parts *and* holding them in three-dimensional alignment on the surface plate. These are items you'll likely only find at a machine tool supply store. If you cannot locate steel, then have a cabinetmaker cut you some precisely square chunks of hardwood (oak or mahogany) scrap to use for alignment blocks.

Cutting Plastic

The only major drawback in working with plastic is that the material is so soft, compared with brass, that it is often a real challenge to hold it while you are cutting it. I've found a swivel-jaw vise, such as the PanaVise, to be the perfect tool to hold plastic while doing heavy filing or sawing. I clamp a scrap of 1" × 4" oak in the jaws of the vise so that about an inch of the hardwood protrudes from the jaws (Fig. 9-2). I can then use the block of wood to support the plastic while I work it with a file or saw. If the plastic needs to be held tightly, I can grasp it firmly between my fingers so my fingers push against the stable support of the vise to hold (squeeze) the plastic to the block of wood. The vise holds the material far enough above the workbench to allow room for the saw blade or file to move. The swivel jaw feature of the vise allows me to tilt the work into a position that is comfortable for both my clamping hand and the hand that is working the file or saw.

I've glued a large sheet of No. 200-grit sandpaper to a 1-foot square scrap of ¾-inch plywood to use as a worktable for flat sheets of plastic. The sandpaper holds the plastic firmly while I mark or score it. I've glued another strip of No. 200-grit sandpaper to the bottom of my steel ruler to make it easier to hold it onto the surface of the plastic when making score lines. I keep some ¼" × ½" strips of balsa wood on hand to cover the jaws of the vise whenever I do have to clamp plastic in the vise (or in the jaws of a lathe collet or chuck). The vise crushes the balsa without damaging the plastic if I'm careful to put just the right amount of pressure on the vise handle.

Using the Score-and-Break Technique

One of the really rewarding aspects of working with plastic is that you can break pieces that are less than .100 inch thick, without expending as much energy as you need to cut

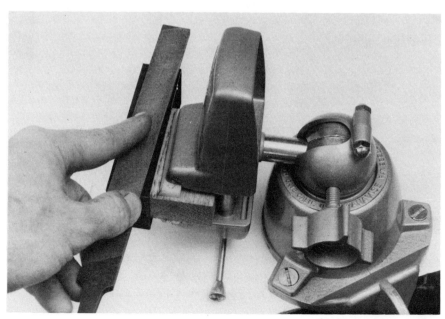

Fig. 9-2 Clamp a scrap of ¼" wood in the PanaVise to support the inside of the plastic car while you file or saw the plastic.

soft balsa wood or paper. The score-and-break technique is essentially the same as that used to cut glass; a light cut or "score" is made precisely on the line where you want the plastic to break (Fig. 9-3). If you hold that cut over a straight and hard corner, you can merely press down on *both* sides of the score, from the top of the sheet (Fig. 9-4), and the plastic will break cleanly along the scored line. You can also use this technique to cut circles or wavy lines. Use a drafting compass with the pen side sharpened so it scores the plastic to score circles. Use a hobby knife with a draftsman's French curve or a flexible metal ruler to make wavy score lines. To break the plastic along those lines, simply work it back and forth between your fingers with your fingers on either side of the scored line (Fig. 9-15). This process is very much like trying to break a wire coat hanger by bending it back and forth until it snaps, but with plastic the material will break along the scored lines. File or sand the broken edge to smooth it.

You can even cut window or door openings this way, but there will probably be a chunk of plastic left in some corners that must be filed. It's easier to gouge, drill, or punch a round hole in the center of the window opening and then whittle the hole to within 1/32 inch of the edge. Finish the opening with a flat file.

The score-and-break method can be used to

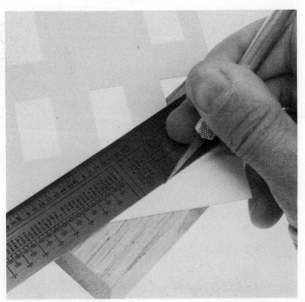

Fig. 9-3 The score-and-break technique, Step 1: Use a steel ruler to guide your knife blade while you make a light cut or "score."

Fig. 9-4 The score-and-break technique, Step 2: Hold a square edge directly beneath the score line and push downward on both sides of the score to "break" the part along the scored line.

speed up the installation of trim strips on models, too; glue an overlength strip of plastic to the model (Fig. 9-5) and, after the glue has dried for a minute, score the strip and break it off over the edge of a block of wood. The special cutting jigs intended for wood, like NorthWest Short Line's "Chopper" (Fig. 10-10), will work just as well with thin strips of plastic.

Surface Textures

The smooth surface of styrene or ABS plastic is perfect as-is for simulating metal or (with clear material) glass. The wood grain on models smaller than 1/64 scale is so fine that it would not be visible on a model, so the plastic looks more like painted wood than real wood on most models. If you want to simulate weather-worn wood in the smaller scales or make wood grain for 1/64 scale and larger models, all that's needed is a steel-bristle wire brush and a hobby knife. Use the blade of the hobby knife to cut some wavy and random slices in the plastic to simulate the few really deep cracks in the "wood." Brush the surface of the plastic with the wire brush using enough pressure to scrape and mark the surface. This must be done before the parts are assembled because it does take quite a bit of pressure. You can substitute coarse-grit sandpaper, and less pressure, or even a coarse-cut

Fig. 9-5 If the trim strips overlap the ends, they can be cemented in place and then trimmed to size using the score-and-break method.

Fig. 9-6 If you work on a steel or glass surface plate with square blocks for vertical support, the walls will certainly be square when the structure is completed.

file for the wire brush if you wish. The rough texture of sand castings that appears on some types of armor plate and on some structural shapes can be simulated by simply brushing on Testors' liquid cement for plastics and dabbing at the surface with the wire-brush bristles as soon as the cement has softened the plastic's surface.

Cementing Plastic

The "magic" of working with plastic is that the cements that are used to attach plastic to plastic dissolve the plastic so that the parts are fused or literally welded together. The seam will be a bit weaker than the surrounding plastic because the solvent leaves a few air pockets behind; the joint looks like a sponge if viewed under a microscope. The fusing action begins only a few seconds after the solvent touches the joint between the two pieces of plastic, and they become "tacky" in about a minute. To the modeler, this means that the model can be assembled almost as fast as he or she can pick up the pieces and attach them to the whole. The solvent-softened joint will not fully harden for at least 24 hours, but the joint is strong enough for careful handling almost immediately.

There are certain dangers to the solvent action, however, including the highly flammable nature of the solvents and their toxicity if inhaled in too large a dose. It's absolutely essential, therefore, to work with plastics in a well-ventilated area away from any open flames. The solvent action can melt the plastic causing the surface to sink or distort if you use too much.

The tube-type cements for plastics have some plastic material dissolved in the solvent that makes them thicker and much slower to dry than liquid cements for plastic. If you use more than a pin-size bead of the tube-type cements, the thick cement can continue to dissolve for *years*. The tube-type cements should only be used on .040-inch or thicker sheet styrene or ABS plastic or on the longer seams of injection-molded plastic kits. Even a thin bead

Fig. 9-7 The small 45 degree notch in the vertical 1 × 2 board prevents the cement from reaching the block so it can be used as an alignment jig.

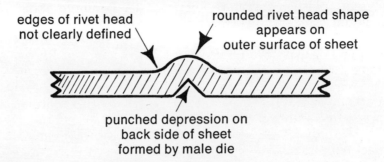
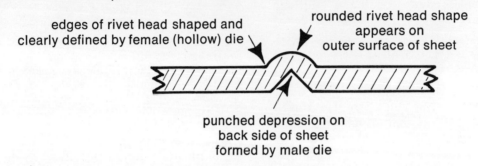
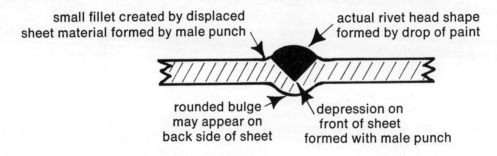
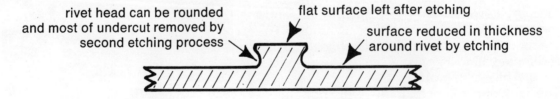
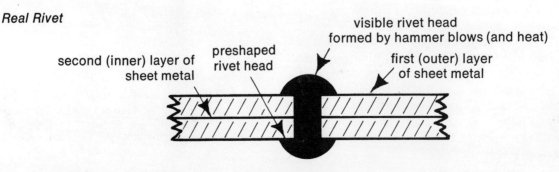

Fig. 9-8 Five cross-section views of methods used to simulate rivets in sheet metal, plastic or card.

of the tube-type cement can cause a sink in most sheet styrene plastic, therefore, you should use the liquid cements for plastics for most assembly work.

A flat surface plate or glass can be used as an assembly table to be sure that parts are aligned properly while being glued together. If you can support one or more walls of a structure, for example, with either steel or hardwood blocks (Fig. 9-6), then the two walls can be held in contact with one another while the liquid cement for plastics is touched to the inside of the corner with a small paint brush or a drafting pen. Capillary action will pull the liquid into the length of the joint. Hold the parts against the surface plate and blocks for about a minute and then assemble the next part. If you must support the inside of the joint (Fig. 9-7), cut one of the corners of the supporting block away at a 45 degree angle so the liquid cement cannot touch the block. The capillary action that pulls the cement into the joints will also pull it toward your fingers if you are touching the seam, as well as toward the surface plate or blocks. If you find the parts are, indeed, cemented to the surface plate or blocks, they can be sliced free quickly with a single-edge razor blade since the solvent action of the cement only works when two plastic surfaces are touching. If you prepaint the parts before assembling them, you must either scrape the paint from the joining surfaces or use 5-minute epoxy or cyanoacrylate cement to hold the painted surfaces together.

If you want to make an assembly jig to hold, for example, the parts of a model fence in alignment while the glue dries, make the jig from wood and coat the wood with a layer of white glue thinned with an equal part of water. When the glue dries, it will seal the wood so the solvent cannot soak in, which makes it easier to lift the parts from the jig after the cement for plastics has dried. The best glues for bonding plastic to wood seem to be some of the water-based contact cements or artist's matte medium—the conventional solvent-based contact cements will melt the plastic. I suggest you reread the previous chapters on assembling the various types of plastic models; there are hints in each chapter that can apply to any type of model you may build.

Embossed Details

Styrene and ABS plastic is soft enough that details can be scribed into the surface to simulate the joints between boards or around the doors in an automobile or truck body. When you scribe the surface, however, a small ridge of plastic is raised on each side of the scribe line. That ridge should be removed by sanding the surface of the plastic with No. 600 sand-

Fig. 9-9 The NorthWest Short Line press with their special riveter table. The knob (*right*) moves the table beneath the punch and die.

paper glued to a perfectly flat block of wood. Even the score-and-break method of cutting styrene leaves a small raised edge on either side of the score line that should be sanded flush with the surface of the plastic. If you merely want to simulate the boards on a wooden structure or railroad car, then you can purchase the "Scribed Siding" sheets of .020 or .040-inch thick styrene from Evergreen. These sheets of plastic are not really scribed; Evergreen uses special milling bits that cut the grooves without leaving the raised edges.

Simulated rivets can be embossed into .015 or .020-inch-thick plastic from the back side of the sheet. If you only need to simulate a few dozen rivets, you can probably get by with pressing the rivets into the back of the sheet with a compass point that has been ground or filed dull enough so it won't punch a hole through the plastic. Place a piece of hardwood beneath the plastic so the compass point won't distort the plastic too much (Fig. 9-8, *top*).

The only way to simulate the true shape of a rivet by embossing is to use both a punch as a male die and a hollow piece of steel as a female die to force the shape of the rivet into the plastic. NorthWest Short Line (Fig. 9-9) and Precision Manufacturing Co. make special punch presses with male and female dies (Fig. 9-10) to shape several sizes of rivets in plastic, cardboard, or soft sheet metal. NorthWest Short Line has a special accessory called The Riveter for their press, which is a sliding table to hold the sheet plastic. The table is moved (as is the plastic) beneath the punch for the rivets by turning a calibrated knob. This allows you to space the rivets very accurately without marking their location with pencil lines and carefully aiming for each "X" every time you emboss a rivet. It would also be possible to use either NorthWest Short Line or Precision rivet punches in a small milling machine like the Machinex 5 or Unimat 3; the NorthWest Short Line Riveter actually moves like a very lightweight milling table. You can get much greater precision by moving the micrometer-calibrated knob to give very precisely spaced rows of rivets than you could hope to obtain by eye.

There are two alternative ways of simulating rivets on thicker plastic or on the surfaces of injection-molded parts too stiff to emboss. The easy way is to simply press a *sharp* compass point into the plastic to make a dent. The edges of the dent will be raised just as they are when you scribe the plastic (Fig. 9-8, *center*). This

Fig. 9-11 The PanaVise press (*left*) with a steel rule die (*right*) and a back-up block of hardwood (*foreground*).

Fig. 9-10 A close-up of the punch and die to show the pin-point size of the male punch with a matching female die below the sheet of plastic.

forms a natural hollow that can be neatly filled with a single dab of thick paint. It is not practical to acid etch plastic to form rivets, although this is a most effective way to simulate them in brass.

The rivets in many of the master patterns for model kits are created by drilling holes in the surface of the material to accept short wire-like pieces of stretched plastic sprue. You can heat the scrap plastic sprue from kits over a candle flame to stretch it into "wires" of any diameter from about .010 inch up. Cut these wires into 1/8-inch lengths and drill holes to match their diameter in the plastic. Insert the wires in the plastic so that at least 1/16 inch protrudes above the surface and touch a drop of liquid cement to each. Let the cement dry for 48 hours; then cut the rivets to within about .010 inch of the surface of the plastic using either a single-edge razor blade or flush-cut diagonal cutters. Brush the still-protruding heads with a brass bristle suede shoe-cleaning brush to round them off slightly. This same technique will work with brass wire and brass surfaces, but the "rivets" must be soldered in place and their heads rounded off with a rotary wire brush in a motor tool or electric drill.

Fig. 9-13 The steel rule die will distort the plastic somewhat around the punched-out window.

Fig. 9-12 The press is used to punch the steel rule die through the sheet of plastic beneath it.

Hole-Punching Techniques

Since sheet plastic can be worked almost like paper, the cutting techniques used on paper often work as well on plastic. An inexpensive paper hole punch, for instance, can be used to start the holes for windows or doors. NorthWest Short Line offers some most helpful male and female punch and die sets for small square and round holes. The punches can be used for instance, to accurately locate the corners of any size window without distorting the plastic. Most of the cutting for pre-cut cardboard kits or packages is done with what are called "steel rule dies." The steel rule die is simply a strip of about .030-inch-thick hardened steel with the edge ground to a knifepoint. A slit is sawed into a block of plywood and the strip of steel (like a steel ruler) is jammed into the slit. The strips of steel can be bent into radii as small as 1/2 inch or used in very short strips to make sawtooth cuts of any shape or size. Every major city has several firms that specialize in steel rule dies; they are easiest to locate by contacting cardboard box companies listed in the telephone book yellow pages under "Package Design & Development."

The steel rule die can be made to cut any size opening you might want in plastic up to about .040 inch thick or in cardboard. The dies will cost at least $30 for a very simple rectangle (like the one in Figs. 9-11, 9-12, and 9-13), so you may want to make one to match your smallest window and use it to "nibble" the corners for larger window and door openings. The steel rule die must be used in a heavy-duty drill press or a modeler's press like the PanaVise unit in Fig. 9-11. Place a block of hardwood between the plastic and the steel rule die and press the die into the plastic (Fig. 9-12). The die will distort the edges of the holes slightly (Fig. 9-13) but they can be flattened out after the material in the hole is pried from inside the steel rule portion of the die. The edges of most windows are covered with strips of framing; if the edges of a die-cut hole are to be exposed, the die should be about 1/32 inch smaller than the hole so you can file the edges a bit larger to remove any distortions.

Heat-Formed Plastics

Most of the plastics used for models will soften to the point of elasticity (where they will stretch permanently) at about the temperature of boiling water. You can use this feature of plastic to stretch scraps of sprue into thin wisps or to make bends in pipes (like the plastic exhaust pipe for a motorcycle in Fig. 9-14). To stretch or bend solid chunks of plastic that are 1/8 inch or more in diameter, hold them just above a candle flame until the plastic is softened. Be careful in this because most plastics *will* burn!! You do *not* want to get the plastic hot enough to melt, just warm enough to bend or stretch. It's better to err on the cool side and have to bring the part back over the flame than to risk the chance of fire or of ruining the part with too much heat.

Sheets of plastic can be stretched over male molds if the four edges of the sheet are clamped in an upper and lower frame like that in the Formicator in Chapter 4. Heat some peanut oil in a baking pan large enough to hold both the frame and the plastic so the oil is about 250° Fahrenheit. Again, be careful because any spilled oil will burn—it's best to heat the oil in an oven with the door open. Remove the pan and place the framed plastic in the oil until the plastic is hot enough so it sags slightly. The plastic can then be pulled over a male mold to produce a duplicate of simple shapes like bubble canopies or domes for a space ship or rounded cowls for an older aircraft. If you want the plastic to capture finer details, like panel lines or rivets, or to conform to complex curves, you must use a vacuum-forming machine to pull it tightly against the pattern as described in Chapter 4. It is *not* necessary to use the oil bath when vacuum forming, however.

Vacuum-Formed Plastics

There are hundreds of aircraft model kits that have been produced using the vacuum-forming process. You can recognize these kits easily because the parts are merely bulges in solid sheets of white styrene plastic. You must cut each of the parts from the styrene *and* fit them together properly to assemble a vacuum-formed kit. There is a lot more work to the process than with an injection-molded kit, but the vacuum-forms are often replicas of fascinating models that are just not available anywhere else. You can even make your own vacuum-formed kits, using the techniques described in Chapter 4, but I strongly suggest you build at least one of the aircraft kits first. You may be able to locate a vacuum-formed kit to build an armored vehicle or an automobile but they are quite rare; most of the vacuum-formed kits are aircraft models.

The vacuum-formed parts can be removed from their sheets using the score-and-break method. Here, the raised surface of the part itself can guide your knife blade while you make a light cut about 1/64 inch from the outline

Fig. 9-14 A candle flame can be used to soften plastic so it can be bent, like this motorcycle exhaust pipe, into curves.

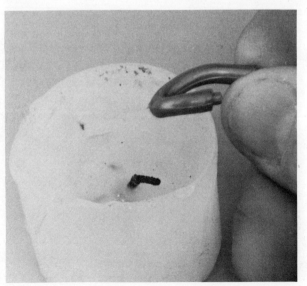

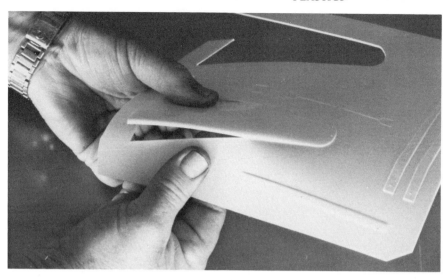

Fig. 9-15 The score-and-break technique works well with vacuum-formed kits; twist the plastic up and down between your hands to break it along the scored lines.

of each part. When you have cut completely around the part, the sheet plastic can be worked up and down in your fingers (Fig. 9-15) until the plastic breaks along the scored line to release the part. You can do the work even quicker than I can describe it, and with only light pressure.

Cement a sheet of No. 400 wet-or-dry sandpaper to a flat sheet of plywood to serve as a sanding block. Wet the surface as you rub the parts across it to produce the perfectly flat joining faces needed to assemble most parts (Fig. 9-16). Glue the No. 400 sandpaper around broomhandles or dowels to make sanding sticks for concave surfaces. The parts can be test-fitted and, if every seam is perfect, they can be assembled using liquid cement for plastics.

Using Filler Putty

The filler putties used to fill gaps between the seams of injection-molded plastic kits are equally useful in filling gaps or making fillets on wood or metal models. Since most putties

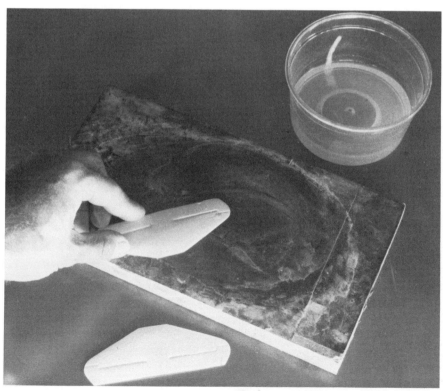

Fig. 9-16 A large, flat sanding block makes it easier to finish the joining surfaces of the parts of vacuum-formed kits.

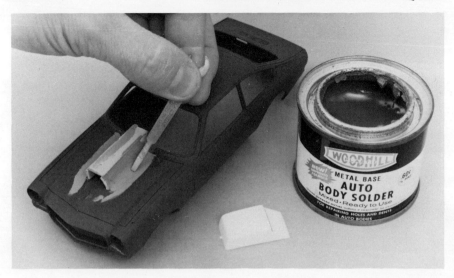

Fig. 9-17 Several brands of automobile body putty are suitable to fill seams or to make fillets on plastic models.

are a plastic material, however, I am covering them in this chapter. Basically, three types of putties are useful for model building: very thin or water putties (like Micro Scale's Quick Silver Putty), pastelike putties (including a wide range of metal and plastic-filled putties intended for use on automobile body repair work), and thicker putties (like the Bondo putty used to fill large dents in automobiles.

I have had no success at all using Testors model putty or Plastic Wood putty; I find both putties to be far too porous for use on a miniature. Most putties shrink or sag as they dry; the thinner or more watery the putty, the more it will shrink. If you have only hairline cracks to fill, a thin putty will work best. If you have a very large area to fill (say, a dent or fillet that is 1/8 inch or more), then the two-part (resin and catalyst) putties such as Bondo are best because they do not shrink as much as others. Be careful, however, when using the thicker putties on plastic because the heat that cures them can melt the plastic or the solvents may attack it.

I've been able to use several brands of automobile "spot" putty for everything; it is thin enough that it can be forced down on the surface to squeeze into hairline cracks. If you need to fill a 1/8-inch sag, simply apply three layers of the putty after allowing each layer to dry overnight. The term "spot" means that the putty is intended only for very small or spot dents. The 3M company makes a product, which the automobile trade calls Green Stuff, that is a suitable spot putty, but I've had equally good results with Woodhill's Metal Base Auto Body Solder (it's really a putty), Duro's Spot Putty, Duratite putty, and Squadron's hobby Green Stuff putties. Simply spread the putty over the crack or seam, using as little as possible, and allow it to dry overnight.

The part can then be sanded with No. 400 or 600 wet-or-dry sandpaper or filed flush with a jeweler's file. Finish the surface with well-wetted No. 600 wet-or-dry sandpaper to remove any scratches. The slightly porous texture of the putty (compared with plastic or metal) may be visible beneath the paint, so it's best to apply either a coat of primer paint or flat-finish paint that can be sanded lightly with wetted No. 600 wet-or-dry sandpaper to match the texture of the puttied area to the rest of the model.

Chapter 10

Wood

WHEN you consider the concept of the master sculptor, you probably envision an artist in a smock hammering away at a chisel, shaping a giagantic block of wood. Wood is the traditional material that we associate with sculptural art forms. In truth, there is no material for a model that has quite the feeling of wood; the material has a surface texture and a heftiness that inspire a favorable feeling.

Wood is seldom the best material to choose when making most models, however, because it doesn't really have the characteristics of a material that must capture shapes and edges a small fraction of the size of the original object. Wood is particularly good as a material to duplicate bridges, fences, or partially destroyed buildings that have a multitude of loose planks. Wood is also an excellent material to use as a pattern or "buck" to cast resin or form sheet metal or sheet plastic.

In the past, the hobby magazines were full of articles on how to build wooden replicas of steam locomotives, automobiles, airplanes, and boats that were made of metal. Those were the days before plastics, however, and before relatively inexpensive lathes and milling machines and detail castings were available to make metal models. It is *extremely* difficult to make wood look like metal because the wood must be very hard to hold a sharp edge (like the edge of a fender well) and this makes the wood extremely difficult to shape. The techniques used for plastic or metal models can also be applied to wooden miniatures. If you're wise, though, you'll use wood only in those applications where it still is a suitable material.

Making Master Patterns

One of the best applications for wood, as a model-making material, is in the production of master patterns that will be duplicated in some other material on the finished model. If the original was wood, as it might be in the case of a boat hull, then wood may also be the best material for the model because the grain can be covered with shellac or other clear paint so it looks like what it is. Even with a ship model, however, you are faced with the problems of making the hull look hollow. A kit may include precut parts for the bulwarks (the sills along the top edges of the hull), but those details can be difficult to add to a scratch-built boat with a hand-carved hull. Most experienced ship modelers seem to prefer to use the hollow hulls made from ribs and planks.

A hand-carved wooden hull can be most useful as the "buck" or master pattern for a fiberglass replica of any of the modern steel or fiberglass hulls. Carve the hull's shape enough *undersize* to allow the use of two or more layers of fiberglass (about .040-inch per layer, per side); protect the hull with several layers of two-part epoxy paint that is sanded smooth, and simply lay the fiberglass cloth and resin over the outside of the wooden hull pattern. When the resin cures or hardens, the hull can be removed. This same process can be used, with equally delightful and lightweight results, to build automobile bodies or aircraft fuselages and wings. Some of the ships and aircraft models in the larger scales have fiberglass parts that are made at the model kit manufacturers.

The wooden master pattern can be used to help shape thin brass, aluminum, or plastic replicas. The brass or aluminum can be shaped by dividing the model into many small panels (about 3" × 4" or so) and shaping each of the panels by hammering them to conform to the contours of the wooden pattern or "buck." The panels can then be joined to make the whole in a manner similar to the way the panels were joined on the prototype. Older warships, for instance, have the steel panels of their hulls riveted to the interior bracing; modern automobiles have their panels welded (soldering would do, on a model) to the inner braces. If you wish to make the model's exterior surfaces from plastic sheet material, the plastic can be heat-formed or vacuum-formed over the wooden buck as described in Chapters 4 and 9.

Carving Shortcuts

Even a professional sculptor does not begin his or her work by whittling away at a square block of wood with a knife or chisel. The profile or "shadow" of the object being carved is traced onto the side of the block, and that profile is sawed from the block with a jig saw as shown in Chapter 4. The profile or shadow 180 degrees away from the first profile is then traced onto the block and cut to produce the major contours of the object, as shown in Fig. 10-2. The curves and contours of the object are then shaped with knife or chisel to within about 1/32 inch of the final shape.

If you're carving a model, you should keep plans to match the size of your model on hand so you can check your progress frequently to be certain you are not removing too much material. When you are completely satisfied with the rough shape of the object, sand the final contours starting with rough-grade sandpaper and progressing to medium and fine grades to obtain a perfectly smooth and perfectly balanced (right-to-left) shape. When you are certain the shape is correct, the interior of the object can be made partially hollow by carving away some of the interior. It's possible to use chisels or gouges to carve inside the model but the work will be much easier if you can use a milling bit in a high-speed motor tool like Dremel's (Fig. 10-5). You must be extremely careful to leave at least 1/8 inch of wood in every area or the model will be too weak.

Supersmooth Finishes

The master pattern that you have carved must be finished to look as smooth as metal even if you do want the wood grain to remain

Fig. 10-1 An electric jig saw, like the Dremel tool, can be extremely helpful in working with wooden models.

WOOD

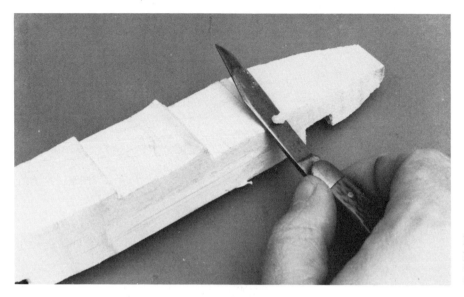

Fig. 10-2 Cut the side and top profiles of the body with the jig saw, then carve the curves into the corners.

visible through layers of clear paint. There are a number of sanding sealers available through hobby shops and furniture finishing stores that will fill in the wood grain without hiding it beneath paint. The sanding sealer is brushed or sprayed onto the wood and, when dry, the sealer is sanded until the wood shows through along the raised portions of the grain. Some of the coarser grains (like balsa wood) may need as many as twelve coats of sanding sealer before the grain is completely hidden. You may be able to save a few coats of sanding sealer if you are going to paint the model. Apply at least two coats of sanding sealer, sanding each one smooth. Apply a third coat and, after it dries, spray the model with one of the automobile primers that are sold in aerosol cans at most auto parts stores. Be sure to test the primer on a scrap of wood sprayed with your sanding sealer because some primers will attack or etch the sanding sealer. The primer is a relatively thick paint that works a bit like sanding sealer except that fewer coats should be needed.

Sand the first two or three coats of primer with No. 400 wet-or-dry sandpaper wetted with water. Sand the final coat or two with wet No. 600 wet-or-dry sandpaper. Several coats of lacquer should give a smoother final finish than either enamel or two-part epoxy paints; the lacquer partially dissolves each of the previous layers to form a single coat of paint. Sand each coat of lacquer with wet No. 600 wet-or-dry sandpaper and polish the final

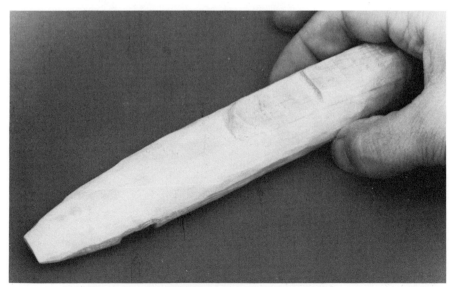

Fig. 10-3 Do not attempt to carve the final shape; leave at least 1/32 inch on each side that can be sanded smooth.

BUILDING MATERIALS AND TECHNIQUES

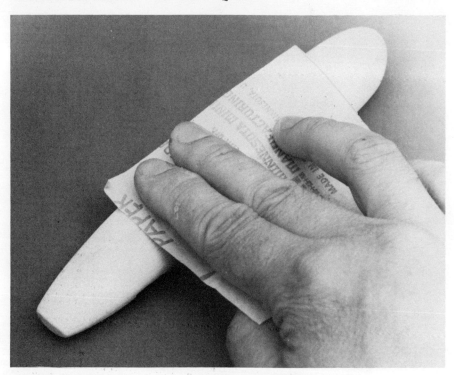

Fig. 10-4 Progress from medium-grit to fine-grit sandpaper as the model approaches its final shape.

coat with rubbing compound followed by a thick coat of wax. The number of coats of lacquer you'll need can vary between four and twelve, depending on how smooth a finish you want.

Fiberglass Finishes

If you are merely going to cover the master pattern with fiberglass, there is no need to completely hide the grain. The surface must be smooth enough so that the hardened resin does not try to "grab" the grain to prevent you from removing the pattern. The wood can simply be sealed with several coats of acrylic enamel or with one or two coats of two-part (catalyst and paint) epoxy paint. The surface of the paint can be sanded with fine sandpaper, followed by wet No. 400 wet-or-dry sandpaper, to remove any remaining traces of the grain. The model must be waxed and coated with a layer of special mold-release compound (from a boat supply store) so the fiberglass resin will not cling to it. You can then apply a coat of resin, followed by the fiberglass cloth and another layer of resin. A second layer of fiberglass cloth and another layer of resin should be added to most boat hulls, but you may want to try just a single layer of cloth for an ultralightweight aircraft fuselage or race car body. Be sure to wear gloves and a respirator when working with the fiberglass cloth because it sheds microscopic slivers of almost indestructible fiberglass. When the resin has cured overnight, the hull can be removed from the wood pattern. You can use that pattern to make replacement hulls (or fuselages or bodies) if you damage the first one! If you want to use your wooden buck or pattern to shape a metal hull or body, the wood should *not* be painted; simply coat it with several layers of wax to help seal the grain.

Fig. 10-5 A Dremel high-speed motor tool, with a small Dremel milling bit, can be used to hollow out the body.

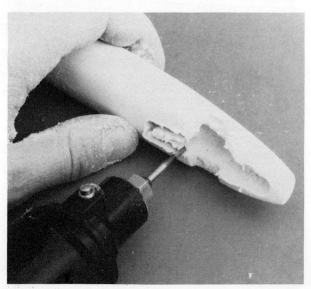

Fig. 10-6 Mount the body on a heavy stand so you can spray on the various coats of sanding sealer and paint.

Lightweight Wood

Wood is still the favorite material for flying model aircraft, although expanded foam plastic and even thin fiberglass are becoming more common in both kits and scratch-built models. Balsa wood is the most suitable of all the lightweight woods for model aircraft. There are several grades of balsa, but most shops only carry the standard grade (relatively soft) and hard balsa. Hard balsa is just a bit softer than pine or fir but not nearly as heavy. The hard balsa is a bit heavier, however, than regular-grade balsa so it should only be used where

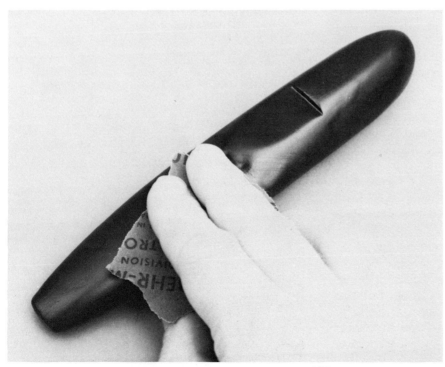

Fig. 10-7 Automobile primer can be used to provide a super-smooth finish. Sand each coat with wetted No. 400 wet-or-dry sandpaper.

hinges or engine mounts are placed to provide extra strength.

Most modelers use thin (1/8-inch) plywood for the cowl formed immediately behind the engine and for the tabs or ribs that actually accept the engine mounting bolts. The plywood is also available at hobby shops that specialize in flying model aircraft supplies. Most of the flying model aircraft are constructed from "sticks" of balsa for the spars that run the length of the fuselage and wings. The airfoil-shaped ribs in the wings, stabilizer (elevator), and rudder are generally cut from 1/16-inch or thicker sheets of balsa. The ribs (bulkheads) that shape the oval or rectangular fuselage are also cut from sheet balsa. Most flying aircraft are covered with special sheets of plastic that is thinner than tissue paper but some of the larger scale models are also covered with sheets of balsa that have been shaped over metal molds or bucks with hot steam and dried. The steam will eventually warp and ruin a wooden pattern but you could certainly use it to shape several models with 1/32- or 1/16-inch-thick balsa wood "skins."

Balsa wood is available in blocks as large as a square foot or more, but it must be ordered in advance and such large blocks are quite expensive. Two-inch by three-inch by three-foot strips of balsa are generally stocked in model aircraft supply stores. These blocks can be laminated into giant blocks the size of a house by gluing the strips together with water-base contact cement. The contact cement will present a bit of a problem when you carve across it, but it's better than most conventional wood glues which dry almost rock hard. For most model aircraft applications, the 1-inch-thick blocks are large enough. Solid blocks are generally used only for the tips of wings, stabilizers, or rudders. The balsa can be shaped with either a jig saw or an electric saber saw clamped to the workbench (Fig. 10-8) and then carved and sanded to the precise profile you desire. It is much easier to carve the compound curves of a wing tip or a cowl than to try to shape such parts as hollow structures.

Duplicating Wooden Structures

Wood is the material of choice when you want to duplicate a prototype that clearly displays individual boards. A fence, wooden bridge, house under construction, or a house that is abandoned and falling apart are a few examples of where the model should be made board by board like the prototype. There is no need for the modeler to use nails to assemble the wooden pieces; nails can be effectively simulated with dots of ink from a drafting pen with a fine (No. 00) tip.

Northeastern offers scribed sheets of basswood that can save a considerable amount of time if you are covering a wall of a relatively

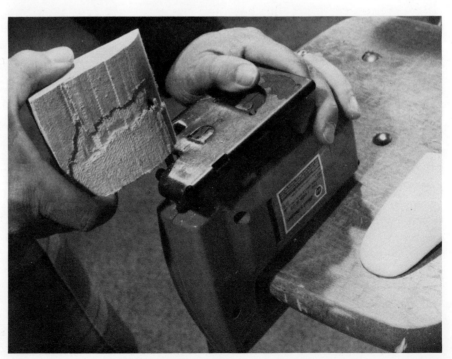

Fig. 10-8 Clamp an electric jig saw firmly to the workbench so it can be used to cut large pieces of soft woods like balsa.

Fig. 10-9 Gordon Johnson's prize-winning abandoned farmhouse diorama features a house built board by board from basswood.

new structure with wood. If you expect to capture the look of wood that has been left unpainted for too long, however, individual strips of basswood are the best and easiest material to use. Balsa wood is *not* a wise choice for a model with any exposed wooden surfaces; the grain in balsa is too coarse to look right on a model in any scale up to one fourth the full size.

If you use sharp blades in your hobby knife, you won't have much more trouble cutting basswood than you would have cutting balsa. The special cutting jigs, like NorthWest Short Line's Chopper (Fig. 10-10), have adjustable stops that make it easier to cut many pieces of equal length. The tool holds an easily replaceable single-edge razor blade in a handle that gives some additional leverage to make the cutting easier. I find I can cut 3/16-inch-thick strips of basswood or 3/8-inch strips by "chopping" through one half of the strip at a time with the tool. An X-Acto razor saw is needed to cut through anything thicker. Sheets of wood can either be cut with a razor saw (if it will reach), a jig saw, or a hobby knife with a very sharp blade. If you use a hobby knife, the secret to cutting thicker wood is to change blades frequently (about one blade per three or four windows is typical); the task is also easier if you guide the blade with a steel ruler and make the cut in several light passes rather than trying to jam the blade through in just one or two cuts.

Learning from Kits

There are some truly incredible structure and bridge kits in 1/160, 1/87, and 1/48 scales available through model railroad hobby shops. There are also some fine building kits and "systems" in 1/12 scale—I've devoted an entire book (*Dollhouses and Dioramas*) to the 1/12 scale systems and kits because there are so many approaches to large-scale structure construction. There are also a variety of approaches to smaller scale construction; some of them are described in Chapters 8 and 9 of this book. I can recommend the Campbell No. 368 "Shed Under Construction" in 1/87 scale for those who want to learn how to build a structure board by board. The kit does not in-

BUILDING MATERIALS AND TECHNIQUES

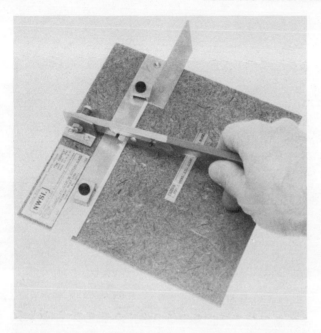

Fig. 10-10 When a large number of pieces of stripwood of identical lengths are needed, the NorthWest Short Line Chopper can cut them just about as fast as you can move the handle.

clude the exterior framing for this particular building, but the shops that sell the kits should also have Northeastern's or Kappler's basswood strips for a clapboard or board-and-batten exterior. The Campbell 1/160 and 1/87 scale bridge and trestle kits are also fine "lessons" in the art of assembling bridges from wood. The Campbell kits, like most kits, include full-scale plans so you can assemble the structure or bridge (Fig. 10-11) directly over the plan. Cement the plan to a scrap of Celotex or similar soft wallboard. Cover the plan with wax paper so the cement doesn't stick to it. The stripwood can then be pinned directly to the plan with map pins, hatpins, tee pins, or straight pins. When the wood is in place, apply a drop of white glue, plastic resin-based glue or Super Jet to each of the joints. When the glue is dry, remove the parts from the plan.

There are a number of limited-production kits offered in the model railroad industry that include incredible detail work with plastic and metal castings for details as small as the hammer (in a workshop) or the insulators on telephone poles. Some of these kits even include special wood items like preassembled ladders, stairs, water tank wrappers, and even preassembled board-by-board wall panels with pre-cut window and door openings. You won't be able to build a more detailed model from scratch than you can from one of these kits. Examine the kits from the Structure Co., Fine Scale Models, Durango Press, Evergreen Hill Designs, Dyna-Models, Detail Associates, S S Ltd./Walthers, and Sequoia Scale Models at one of the larger model railroad hobby shops. You can learn some true contest-winning secrets from the techniques, the instructions, and the actual assembly of kits like these.

Staining Wood

If you are assembling a board-by-board replica of some structure or bridge from genuine

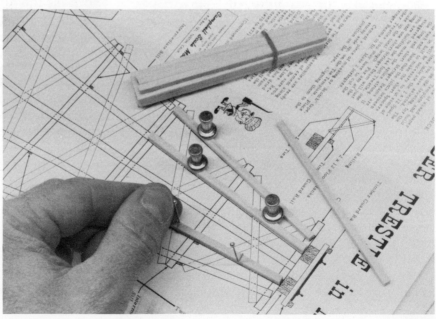

Fig. 10-11 Cover the full-size templates with wax paper and then pin the parts directly to the plan and cement the cross braces in place.

Fig. 10-12 Each of these trestle bents was made in a jig. Another jig was used to align them. Only the diagonal braces are added "freehand."

wood parts, it is best to paint or stain the individual pieces *before* they are assembled. One of the purposes of using wood on these types of miniatures is to allow the grain and texture of the wood to be visible on the completed model. If you wait until the model is complete to stain or paint it, the slightest glue smear or even soaked-in glue near the joints will be a slightly different shade from the rest of the model. If you are going to paint the model as though it were almost new, the paint can be be applied to the completed structure as described in Chapter 8.

You can use almost anything to stain or color the wood *except* paint or thinner with an oil base. The oil base will prevent the glues from adhering properly. Alcohol, paint thinner, or even water-thinned paints or stains are fine. Shoe dye, Rit clothing dye, acrylic artist's colors (thinned in water), Floquil's stains for models, and most of the stains for full-size picture frames all work well on models. The only trick is to carefully match the actual colors on the prototype you are modeling—using "oak" stain to try to simulate weathered oak won't even come close to the actual color of the wood on the real thing. Most woods age to a gray-green color even when creosoted. It's the subtle differences in the shades that make the difference between a model and a toy, however, so you'll have to examine the actual structure you are modeling and match your stains to it.

The stains can be applied to long strips of wood before they are cut by wiping them with a rag soaked in the stain. If you are going to stain individual pieces after they are cut, then it's best to use a large wire mesh strainer to dip the parts into a baking pan filled with the stain. Spread the stained parts on old newspapers to dry. If you are staining individual parts, be sure to keep them separated by length during the staining process. If you are staining the stripwood before cutting it to length, experiment with your stain to see if it should be a bit thinner or a bit thicker to match the stain on the sides to the stain on the cut ends when you touch up the finished model. If you want to simulate peeling paint, prestain the parts and let the stain dry, then coat the areas where you want the paint to peel with a thin layer of rubber cement (from a stationery store) and let it dry. Finally, paint the surface and, when that paint is dry, carefully pick it off with a sharp knife and tweezers to duplicate the rolled-edges and raised chunks of chipped paint.

Chapter 11
Metal

THE most valuable feature of metal, to the modeler, is the material's ability to be worked into very crisp and extremely precise shapes. If a single rivet on a steam locomotive tender, on an old warship hull, or on a tank is out of line from all the others by just an inch it will be visible on the full-size vehicle. On a model of that vehicle, the same rivet cannot be out of line more than a few *thousandths* of an inch or it will appear as gross and sloppy as that inch-out-of-line rivet on the prototype. When you reduce the proportions of the prototype by 1/48 or 1/87 or whatever the scale you have chosen may dictate, you are also reducing any margin for error or misalignment by that amount.

Metal is also quite hard and therefore requires a considerable amount of either physical energy or money (to buy machine tools) to shape it. Today, it's foolish to try to shape metal with files or to turn round parts by chucking them in an electric drill and shaping them with a file. If you must limit your expenditures on tools, then use a softer material like plastic that can be worked with hand tools. There are several small lathes, milling machines, and drill presses (mostly in machines, that combine all three functions through the use of accessories) that are perfect for creating models from metal. The etching process is also relatively expensive but a truly sensible way to obtain precision shapes in sheet metal. Even though soldering guns, torches, and carbon electrode systems are expensive they are extremely helpful. The powered machine and modern tools make it as easy to obtain precision in metal as it is to shape wood or plastic with hand tools. The hand tools needed for work with other materials are also helpful for fine shaping, assembly, and detail work with metals. The precision measuring tools, surface plate, and alignment tools discussed in Chapter 9 are just as important when working with metal as with plastic.

Suggested hand tools for metal working are (clockwise, from upper right, as shown in Fig. 11-2):

Bernz-O-Matic propane soldering torch
Wood clothespins
Self-clamping aluminum tweezers
Weller 100 to 140-watt soldering gun
Sta-Brite Stay Clean acid soldering flux
Sta-Brite Silver-bearing solder
Ersin rosin-core 60.40 solder
Number-size drill bits from 1 through 60
Number-size drill bits from 61 through 80
General Tool pin vise
File card (wire brush cleaner for files)
Wood handle to hold jeweler's files
Round (rat tail) jeweler's file
Rectangular jeweler's file
Triangular fine-cut small file
Rectangular medium-cut mill file
Small screwdriver for slotted-head screws
Small screwdriver for Phillips screws
Jeweler's medium-size screwdriver
Conventional pliers

Fig. 11-1 The steel surface plate and rectangular alignment blocks can be used when cyanoacrylate cement will hold metal parts together.

Needlenose pliers with serrated jaws
Needlenose pliers with round jaws
Needlenose pliers with flat (smooth) jaws
X-Acto hobby knife with spare blades
Flush-cut diagonal cutters
Small diagonal cutters
Jeweler's saw with spare blades
X-Acto razor saw with handle
Tin snips

Brass and Nickel Silver

There are really only two metals worth considering for the construction of most models: brass and nickel silver. Nickel silver is really just a different alloy of copper and nickel from the alloy used for brass; the proportion of nickel is increased so the material has more of a silver appearance than the goldlike tint of brass. Bronze is also suitable for bar stock but it is often a bit too hard and it can contain air pockets or particles of sand that can make it difficult to work with.

Nickel silver is much more expensive than brass and more difficult to locate; it's really most useful when you want to simulate a bare metal surface that looks something like steel. Model railroaders, for example, prefer nickel silver to all other materials for the rail used in their track. Both brass and nickel silver are available in sheets ranging from "shim stock" that is .002 to .005 inches thick to sheets about .5-inch thick.

Most model railroad shops carry the K & S brand line of 4" × 10" brass in .005, .010, .015, and .032-inch thicknesses. The K & S assortment of metals includes .008 tin, .016, .032, and .064 aluminum, and .025-inch-thick copper sheets as well as a wide range of brass tubes in square, round, and hexagon shapes, solid brass rod, round aluminum tube, brass angles and channels and strips and hardened steel piano wire. Mail order supply firms like Caltronics, Special Shapes, Milled Shapes, Craftsman Specialty Supply, and Walthers can provide a variety of other shapes and sizes of sheet, tube, and bar stock in brass as well as in such "exotic" materials as nickel silver, stainless steel, and copper. Steel is very difficult to solder and aluminum is almost impossible to solder. Tin is suitable but much more difficult to work than brass. The only metal you might consider as an alternative to brass or nickel silver for model work might be aluminum and, then, only for hand-formed panels like race car hoods or aircraft panels that do not need to be soldered in place.

Cutting Sheet Stock

The thin sheets of brass can be cut with heavy-duty scissors or with tin snips; since either method will curl the edge of the sheet, however, the cut must be about 1/64 inch outside the final line so that you can bend and file the edge to the correct size. Radio supply stores sell a small tool called a Nibbler that can cut holes in the thin metal used for radio chassis and cabinets. The Nibbler (Fig. 11-3, *top*) works on the shear principle to cut approximately ¼" × 15⁄16" chunks of metal from the part. It leaves a slightly rough edge that must be filed flush, but it does not distort the metal; it's perfect for cutting windows because you can work it into a ¼-inch diameter pilot hole in the center of the window to nibble out the straight edges. Brookstone Tool Company specializes in unusual tools; they import a sheet metal cutter (Fig. 11-3, *bottom*) that also works on the squeeze principle to use leverage to multiply the strength of your grip. The sheet metal cutter also works on the shear principle but it cuts a continuous strip of metal about 3⁄16 inch wide from the sheet. The edges of the cut are slightly ragged so they also must be filed smooth. Either tool will cut up to .050-inch-thick sheet metal (or up to .080-inch-thick sheet plastic) with little more effort than is needed to squeeze a paper punch. Either tool can also be turned, as it cuts, to make curves and other irregularly shaped cuts.

Sheet metal thicker than .050 inch or so must be cut with a saw of some type. For very thin cuts, there's nothing better than a jeweler's saw, although a hand-held jig saw with a fine-tooth blade will work almost as well with a wider cut. Metal-cutting blades are available for electric jig saws and saber saws like those discussed in Chapter 10. The metal must be moved very, very slowly into the teeth to prevent jamming in almost any power saw. Do not try to cut metal with a saw blade that has the coarse teeth needed to cut wood. The high-speed motor tools (like Dremel and Foredom) are quite useful for shaping and cutting metal of all thicknesses. Cutoff disks made of the same material as grinding stones, can cut 1⁄32-inch-wide slits through metal. The disk's diameter is smaller than that of the motor tool, however, so you must either make the cut near the edge of the sheet metal or angle the cut so the cutoff disk slices the metal with the body of the motor tool rubbing across the surface of the metal. The high-speed milling and

Fig. 11-2 A suggested minimum assortment of hand tools to build metal models from kits or from scratch (see page 144 for names).

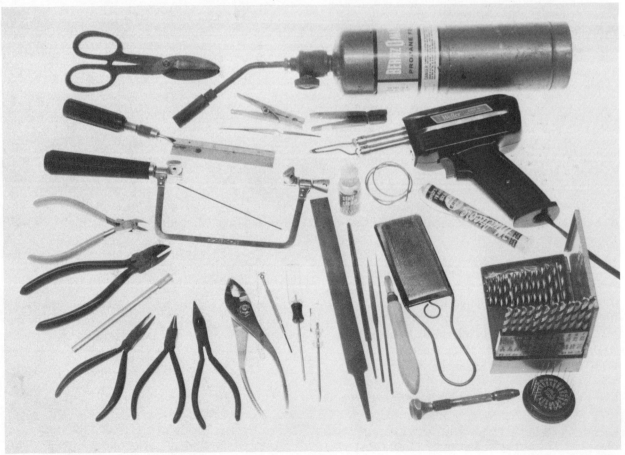

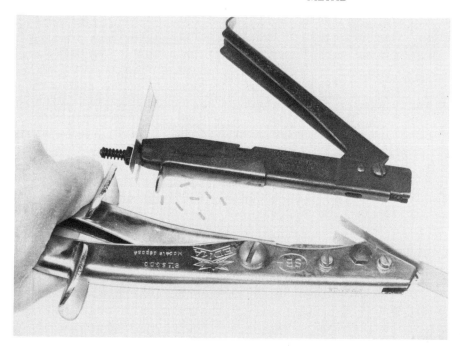

Fig. 11-3 The electronic hobbyist's Nibbler (*top*) cuts tiny rectangles of metal with each pull; the larger sheet metal cutter, from Brookstone, cuts a continuous curl of metal.

grinding bits that are available from Dremel can also be used to carve metal in any thickness.

The most time-consuming portion of any project in metal is the final shaping and fitting. This type of work is usually done with files, but you can speed up the process and reduce the amount of hard labor considerably by using these high-speed grinders to trim the rough-cut edge of sheet metal, for instance, to within .010 inch of the proper size using a high-speed grinding bit; finish that final fraction (and remove any burrs left by the grinder) with files. Straight edges and convex curves can be shaped even easier using a belt grinder like those sold by Dremel (Fig. 11-4), Sears, and Rockwell. If you must budget your tool purchases, I would rank one of these grinders as second only to an electric drill in the selection of power tools.

Etching Sheet Metal

The process of removing metal by literally dissolving it in acid is called "chemical milling" by the professionals and etching or photo-etching by modelers and the printing trade. It is possible to set up the process in your workshop but the costs are so high that you would really need to sell kits made with this process to justify doing the work yourself. You can, however, save a major portion of the cost of etching by preparing your own artwork like that shown in Chapter 7. I suggest you

Fig. 11-4 A belt grinder, like Dremel's, removes most of the tedious labor of hand-filing from any metal-fitting project.

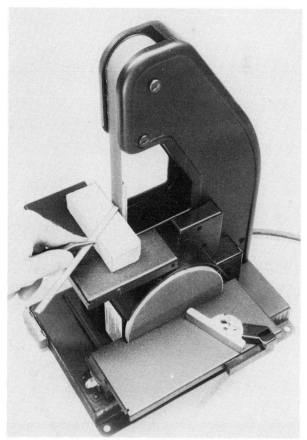

Step 1—Metal Ready for Etching

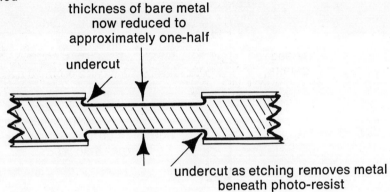

Step 2—Metal Partially Etched

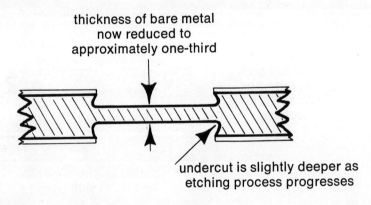

Step 3—Metal Nearly Etched Through

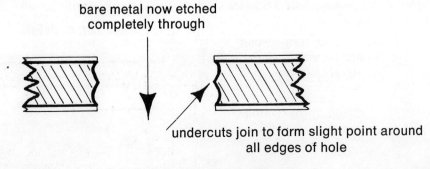

Step 4—Metal Etched Through

Fig. 11-5 The four stages of the photo-etching or chemical-milling process.

Method A—Flat Metal after Etching

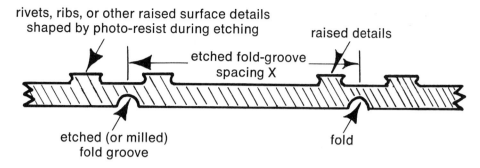

Method A—Bent Metal Corners

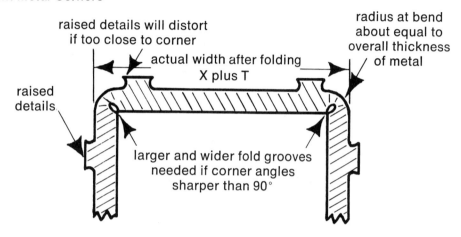

Method B—Reinforced Square Corner

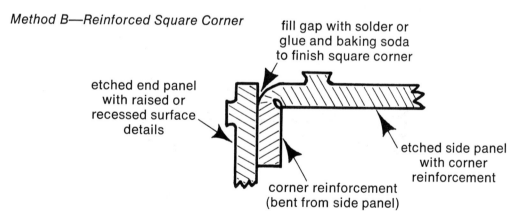

Method C—Reinforced Square Corner with Raised Detail

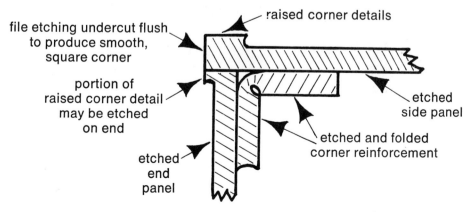

Fig. 11-6 Three methods of providing for accurate corners when designing parts to be produced by etching.

consider using the etching processes to cut sheet metal only if you have at least three identical parts to make where there are a number of holes (like a locomotive cab) or complex surface details like rivets or door outlines. If you only need one or two parts, it's almost as easy to cut each hole or to emboss each rivet (see Figs. 9-8 and 9-10) as it is to prepare the artwork for etching.

The lines that you draw to indicate the edges of windows or the surfaces of rivet heads will be photographed by the chemical milling firm so that they appear clear on the otherwise black film negative. The surface of the brass sheet stock is coated with a special light-sensitive photographic chemical called a "photo resist." When the artwork's negative is placed over the coated sheet of metal, light is directed toward the surface. The light deactivates the photo resist that is beneath the clear areas of the negative, leaving the photo resist only *between* the lines that indicate the edges of the door or window openings (Fig. 11-5, *top*). The sheet metal is then suspended in a small vat of acid. The acid will slowly dissolve the metal from any areas where the photo resist has been deactivated (Fig. 11-5, *bottom*) to cut neatly along the lines indicated on your original artwork. This is how a simple cutout cab like the one if Figs. 7-14 through 7-19 is made.

The chemical milling or photo-etching process can also be used to etch away only a portion of the surface of the metal if the process is stopped (removing the metal from the acid bath) before the acid eats all the way through the metal (Fig. 11-5, *center*). This is how raised lines and details like rivet heads are produced (Figs. 7-9 through 7-12 and Fig. 9-8). This form of the process requires *two* negatives; one with the edges of all the openings to be etched clear through *and* the rivet or other "raised" details, and a second negative with just the lines that are to be etched clear through the sheet. The first negative is positioned over the front of the metal and photo resist and the second negative is positioned over the back of the sheet of photo resist-coated metal. The metal is then etched from both sides at the same time. With the single-side etching process, the entire back side of the metal was coated with a compound or tape to protect it from the photo resist. The two-sided etching process is needed when you

Fig. 11-7 The Greenhalgh 1/160 scale Model T automobile kit is an example of the clever use of etched-brass parts.

need both cut-through lines and raised surface details. The most common thicknesses of material for brass are between .010 and .012 inches with about .003 to .004 inches of that etched away for any "raised" details such as rivet heads. The two-sided etching process can also be used to make "fold" lines on the backsides of the parts by etching those lines only part of the way (about .003 to .004 inches) through the metal (Fig. 11-5). The firms that make kits using photo-etched parts are often extremely clever in their use of this process. The 1/160 scale Model T kit (Figs. 11-7 and 11-8) from M. Greenhalgh Co. (3623 S. Yampa Way, Denver, CO 80013) uses fold-up etched parts for the body with castings only for the round-surfaced tires and seats. The model railroad freight car kits from Quality Craft also include some clever uses of etchings. If there is no printer or chemical milling firm near you, firms like Fotocut (Box 120, Erieville, NY 13061) can do the work by mail.

Fig. 11-8 The thumbnail-size model would, indeed, be difficult to shape by hand; the etching process allows the parts to be designed several times their final size and reduced by photographic processes.

Machining Metal

When you remove metal with a machine that allows you to control the amount of metal being removed the process is called "machining." Drilling is a common type of machining process. When the part you are making is round, like a dome or bell, the round bar stock may be chucked in a lathe and turned against a cutting bit that is held stationary in a tool holder in the cross slide of the lathe. The cross slide can be moved across the axis of the lathe by turning a crank which moves the cross slide on a screw mechanism. The cross slide can also be moved along the axis of the lathe with a similar crank or wheel and screw thread. These two screw-thread adjustments provide the precision in working with a lathe or a milling machine because the collars around the crank handles are marked in thousandths of an inch. You can, then, remove thousandths of an inch of metal by turning either crank so the cutting bit moves closer to the revolving bar of metal. There is no need to measure and mark the amount of the cut to try to match that mark; with a lathe or milling machine the crank allows you to literally "dial-in" how much to cut. For example, to remove exactly .302 inch from a round bar (to reduce its diameter by .604 inch), you would crank the cross slide until the cutting bit just touched the spinning bar stock. Set the adjustable collar on the crank to zero inches and gently rotate the crank to remove a thin sliver of metal each time the bar revolves until the mark on the crank lines up with .302 inch on the adjustable collar. It would be wise to stop the machine two or three times during the process to double check the crank (and your math) by measuring the partially cut diameter with a micrometer.

Milling Machines

The Machinex 5 from Edestaal (Figs. 11-9 and 11-13), the Unimat 3 from Emco-Lux, and the Sherline lathe can all be converted or adapted to small vertical milling machines. A lathe operates by rotating the part for most of the common operations while a milling machines operates by moving a stationary part into a spinning cutting bit. Both machines use virtually identical cranks with micrometer-precise markings which allow you to control the depth of the cut to less than a thousandth of an inch. Most milling machines position the motor and cutting bit on a vertical shaft so that either the unit or the work that is being cut can be moved up and down as well as back and forth and across the face of the cutting bit. That third dimension of adjustment makes the milling machine a far more useful tool, to most modelers, than a simple lathe. It is possible to chuck the work in the head of most milling machines sold for modelers (rather than the milling bit) so the milling machine can be used like a "vertical lathe." The important feature that makes the milling machine so useful is

BUILDING MATERIALS AND TECHNIQUES

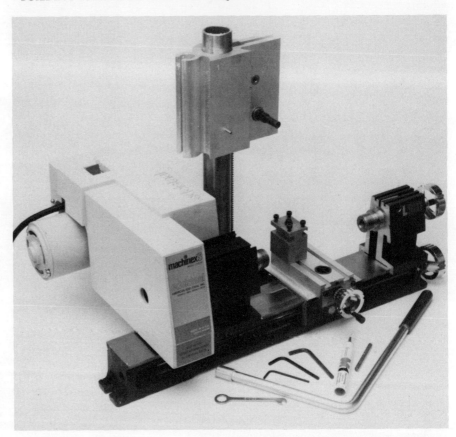

Fig. 11-9 The Machinex 5 can be used (as shown) as a lathe or the drive unit can be mounted on the uppermost bracket to convert the tool into a drill press or milling machine.

a crank with micrometer adjustment like that in the upper left of the Machinex 5 illustrated in Fig. 11-13. The crank on this "vertical feed" accessory controls the up and down movement of the cutting head with micrometer precision.

The milling bits or cutters will shake the part rather violently as they dig into it, even if the part is moved into the cutting teeth very slowly. The part, therefore, must be clamped firmly to the machine using a special milling table that has keyways to accept hardened tee bolts and nuts. The same types of clamps, vee blocks, and step blocks that are sold for larger professional milling machines are available in small sizes to fit the machines used in model work. You may be able to locate these clamping parts at a local machine tool supply store; if not, they are available from Carr Lane Manufacturing Co. (4200 Carr Lane Ct., Saint Louis, MO 63119). You can use the milling bits sold by Dremel or their cutoff wheels (Fig. 11-13) mounted in a drill chuck. For really precise work, however, a special collet is needed to grip milling bits in the machine. A different collet will be needed for each size milling bit. The collets are available from the same firm

that makes the milling machine, but the bits must come from an industrial machine tool supply firm. Caltronic Labs has a few smaller size milling bits that are most suitable for work with models. Each of the lathe and milling machine manufacturers offers instruction booklets that describe the details of the turning or milling processes and there is no point in summarizing those publications here. I suggest you purchase the booklets from each of the makers to help you decide which brand and type of machine is best for your model building requirements.

Assembling Metal Parts

Most modelers think that metal parts must be assembled by soldering the parts together. Today, you don't really need to know how to solder to build a model from metal pieces. The conventional (watery) cyanoacrylate cements and the thicker ones (like Goldberg's Super Jet) can be used to assemble smaller parts and etchings as described in Chapter 7. Larger parts can be assembled using small screws and nuts, with the smaller parts that add only details to the model held in place with cyanoac-

Fig. 11-10 The round brass bar is held in the lathe's three-jaw chuck (*left*) which revolves. The cutting tool is held stationary in the tool post (*right*).

rylate cement. Solid parts such as domes or whistles can be drilled and threads for a screw cut inside the hole (a process called "tapping") as shown in Fig. 11-11. It is not necessary to drill the hole or to tap it in a lathe; both processes can be done by hand using an electric drill to make the hole and a pin vise (Fig. 7-7) to hold the tap while it is rotated into the hole. The most commonly used taps in America are the 00-90 (about the size of the hinge screws in eye glasses), 00-80, 1-72, and 2-56 sizes. Walthers and Caltronic can supply the drills, taps, dies (to cut external threads), screws, nuts, and washers in these small modeler's sizes. Larger sizes are usually available at hardware stores.

The tap must be twisted into the hole you wish to thread until you can feel that it is binding or locking up. Back the tap out (by turning it counterclockwise) to help clear away the chips of metal, and twist it back into the hole once again. The tapping operation consists of a series of these in-and-out movements. Try to drill the hole completely through the part to leave room for the almost ¼ inch of taper on the tips of most taps. Special "bottoming" taps are available to thread that last ¼ inch of a "blind" hole, but they are easily broken in such small sizes. A light machine oil (like 3-in-1 or sewing machine oil) will provide enough lubrication to prevent a conventional tap from being easily broken.

Fig. 11-11 The lathe can be used to tap holes precisely. The same setting used to drill the hole is used to tap it except that the three-jaw chuck is turned by hand for the tapping operation.

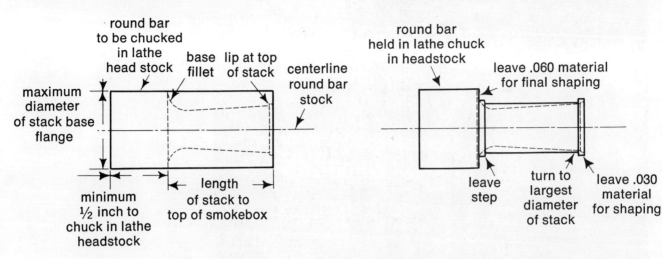
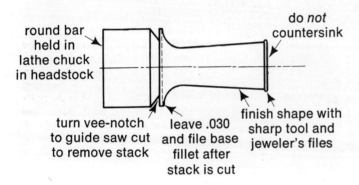
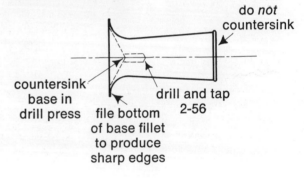
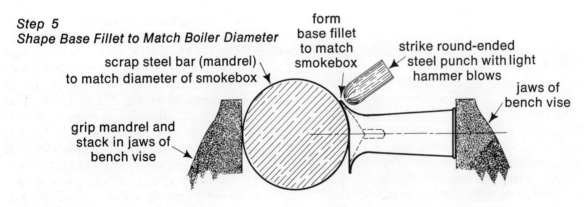

Fig. 11-12 The six stages needed to turn a locomotive stack from a solid round bar and to flare its bottom end to fit around the locomotive boiler's smokebox.

Fig. 11-13 The Machinex 5 with the drive unit mounted vertically and a special micrometer-feed attachment fitted to allow precise vertical control for milling. The black plastic part is clamped to the milling table with tee-bolts (one is on the bed), bars, and Carr Lane brand step blocks. A Dremel cutoff disk is chucked in the tool to remove material from a plastic tender body.

Soldering

The most common method of assembling metal parts has been to join them with a molten metal bond of a tin and lead alloy called solder. The process of melting the tin and lead solder to join the parts is called "soldering." The molten material finds its way into the molecular structure of the metals that are being soldered so that, if the joint does break, it will break along the solder itself leaving a coating of solder on both joining surfaces. The process differs from welding, which involves heating the joining surfaces to the point where the metal itself melts; the welding rod used in the process merely acts as a filler. There should be no joint on a model that needs to be so strong that welding is required. If high strength is needed, as it might be on the frame of a locomotive model, then use screws and nuts to make the joint. The solder can be melted with either a hot soldering iron or a soldering gun (Fig. 11-2) with the flame from a special soldering torch (Fig. 11-2), or with a relatively new process called "resistance soldering" (Fig. 11-15) which uses electrical current flowing through the metal to melt the solder. The solders used for most model work melt in the range of about 275 to 425°F. The basic principles of soldering are the same regardless of which tool is used to melt the solder or which type of solder is used. Solder will not adhere to any metal surface that is greasy, oily, dirty, or even oxidized. The joining surfaces must be perfectly clean, which means that they must be filed, scraped, or wire brushed just a few minutes before the solder is melted. Oxides can form in a matter of minutes after the metal is cleaned, and this can prevent the solder from adhering properly. Light coatings of oxide like this, however, can be removed by using acid flux just before the solder is applied.

Some kind of simple assembly jig will be needed to hold the parts being soldered because they'll get too hot to be held by hand.

It's best not to use pliers or a vise because the heavy metal will act as a heat sink to draw so much of the heat away that the solder will not melt. Wood clothespins, aluminum tweezers, and blocks of hardwood are the best tools to use to position the parts where you want them (Fig. 11-14).

Clean the joint and apply a drop or two of acid flux to the surfaces being joined and then spread the flux over them with a scrap of wood. Dilute hydrochloric acid is available at most hardware stores for this purpose or you can use the Stay Clean flux from Stay-Brite or the Tix flux from Brookstone Tool. Keep some water handy to flush your fingers if they contact the acid and be sure to keep the acid away from children. Wear goggles and a respirator, if you are working near it, because the vapors of the evaporating acid can burn you.

The rosin flux should be used *only* for electrical connections where the flow of electricity, not strength, is the most important consideration. I prefer the Ersin brand rosin core solder for electrical work. For strength, however, always use solid-wire solder. The soldering iron, gun, torch, pencil, or resistance rig is used to heat the parts, *not* the solder. Touch the solder to the parts (*not* to the soldering instrument) until the metal is hot enough to melt the solder. The solder will displace the acid flux as it evaporates from the heat. I keep a wad of damp facial tissue handy to cool the solder joint immediately. When you are through soldering for the day, wash any remaining acid flux from all of the parts by flushing them with water under a faucet.

Soldering Tools

The fourth "secret" to soldering (in addition to clean surfaces, the use of flux, and the need to heat the metal rather than the solder) is the use of a soldering instrument or tool that produces ample heat. The 15 to 55-watt soldering irons, pencils, and guns that are most commonly available are really only suitable for soldering electrical wire connections or very small parts. A soldering iron is like a wood-burning pencil in that it remains hot whenever it is plugged in. A soldering gun (Fig. 11-2) is only hot when you pull the trigger. The Weller 100 to 140-watt soldering gun produces enough heat on either setting for most of the work where a soldering iron is useful. Some modelers, however, prefer the larger heat sink capacity of a 150-watt or 200-watt soldering gun or iron for larger parts. I feel that any part large enough to draw heat that quickly is best soldered with either a soldering torch or a resistance soldering instrument (Fig. 11-15).

The soldering torch is used very much like a soldering iron except that the flame is directed to the joint with the solder held far enough away from the flame so the metal will melt the solder. If you find the Bernz-O-Matic

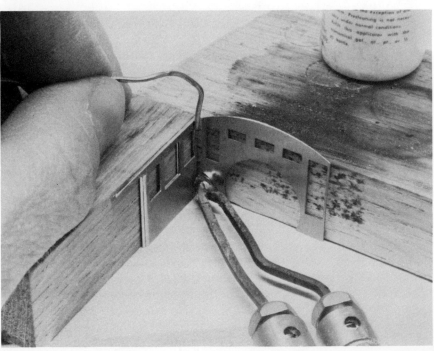

Fig. 11-14 Hardward blocks should be used to make temporary jigs to hold parts that are being soldered together. Heat the metal, *not* the solder itself, with the tip of the soldering gun.

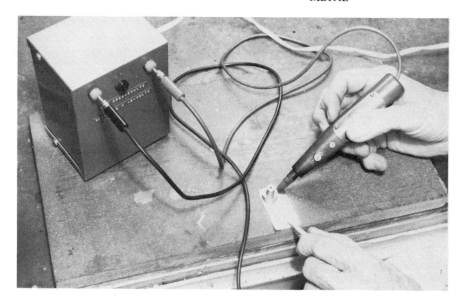

Fig. 11-15 This homemade carbon electrode (resistance soldering) outfit has a probe (*right*) with a push-button on–off switch.

propane cylinders awkward to hold, more expensive cartridge-type units are available, such as the Microflame torch, that burn a mixture of butane and an oxidizing gas (similar to oxygen). The resistance soldering outfits, like the homemade unit in Fig. 11-15, are, in my opinion, the soldering equivalent of the "instant" cyanoacrylate cements—the quick and easy way to perform a rather tricky job. Some of the resistance soldering outfits include a pair of tweezers to clamp the parts together; others simply have a carbon-tipped probe. Some have a foot switch so the unit can be turned on and off with the probe or tweezers holding the part in place until the solder hardens; others have a push button on the handle that serves the same purpose. Resistance-soldering outfits have a transformer that reduces the voltage to a relatively harmless 24 volts or so, but the temperature at the tips of the tweezers or probe can exceed 1000°F! Resistance-soldering rigs are relatively expensive—between $75 and $200—but they make soldering far easier and quicker.

Sweat Soldering

There are two ways to apply the solder to the joint between the two metals: by holding the parts together and touching the solder to the joint so a combination of the evaporating acid flux and capillary action pulls the solder into the joint (Fig. 11-14) and by "sweat soldering." The sweat soldering method requires the surface of at least one of the parts to be heated and coated with a thin layer of solder. When the solder cools, acid flux is applied to both parts and the two are held together while the soldering instrument is applied a second time. The heat from the soldering instrument melts the solder, bonding the two pieces together. The process leaves the hand that would normally hold the solder free to hold the wood or clothespins to position the parts. If you are using a soldering gun or iron or a torch, the tool must be removed from the part so the part can cool enough for the solder to solidify. With the resistance-soldering system, however, the probe or tweezers that are in contact with the part can be left in place; simply turn off the electrical switch and the heat stops immediately. You still have to wait for the heat that is retained in the metal parts themselves to dissipate enough for the solder to harden. The resistance-soldering outfits work on the principle that electricity flows through the probe or tweezers, on through the two parts being soldered, and back to the transformer. A second wire is, then, attached to one of the parts with a simple spring-loaded clip. The clip will get almost as hot as the probe, so it must be positioned away from solder joints. Most of the heat is concentrated directly under the probe or tweezers, however, so you can solder several parts to the same piece of brass without worrying about any of the previous solder joints melting as you add more parts.

Multipart Soldering

One of the major problems with solder as an assembly medium is that the previously soldered joints can easily be softened while you are adding another part to the assembly. A steam locomotive boiler, for example, may have dozens of solder joints to hold all its parts in place. Some of those parts can, of course, be attached with screws or pins so soldering is not required. This is really the best method for large brass parts such as solid domes. The resistance-soldering outfits allow you to solder very close to a previously soldered joint without melting the solder at that second joint. Even those outfits can sometimes melt the joints between smaller parts, however.

The best system to use when several parts are to be soldered in sequence is solders with different melting points. Silver solder is used by some welding shops but its 1145°F melting point requires a very hot welding torch; it can be used to attach very large parts if you can persuade a welding shop to do the work for you. The Stay Brite Silver-Bearing Solder, from J. W. Harris has a melting point of 425°F; most 50/50 (tin/lead) wire solders melt at 325°F; 60/40 (tin/lead) solders melt at about 300°F, and the Tix-brand solder sold by Brookstone Tool Co. melts at about 275°F. The trick is to solder the larger parts with Stay-Brite, the next smaller parts with 50/50 solder, medium-size parts with 60/40 solder, and tiny parts with Tix solder; the smallest parts can then be attached with cyanoacrylate cement. If you hold the soldering instrument to the parts just long enough for the solder to melt, you should be able to remove the heat quickly enough to avoid softening the solder on the next smaller part (that has been soldered with solder of the next *higher* melting point). If two parts are close together either they can be soldered with solders of different melting points or the first solder joint can be protected with a heat sink made of a wad of wet tissue or a metal clamp that will pull the heat away from the joint before it can melt the solder.

The Last Tricks

Your solder joints will be successful every time if you remember a couple of other tricks: keep the tip of the iron "tinned" by rubbing off the crusty accumulation of oxides on a block of sal ammoniac (large hardware stores sell them). Wipe the tip on the sal ammoniac then touch it to the solder to cover the tip with a shiny thin coating of solder. If the tip becomes pitted, let it cool, file it to remove the pits, and then retin. Soldering iron and gun tips that are plated with nickel–iron will last far longer than bare copper tips before they become pitted.

When you apply the solder, use as little as possible. The real experts never touch the model with the end of a piece of wire solder; they slice the wire into tiny disks of solder with a hobby knife. For very small work, those disks can be pounded flat and cut into pepper-size dots of solder.

Coat the surfaces to be soldered with acid flux, and then place one of the disks or dots of solder in place (the flux should hold it there). The second part can usually be positioned next or you can heat the area around the disk or dot of solder to allow the use of the sweat soldering technique. The sweat-soldering method is one way to keep excess solder to a minimum, but it takes a bit more time unless you're using a resistance-soldering outfit. When all of the soldering is completed, scrub the model thoroughly with kitchen cleanser to remove any traces of acid flux and to roughen the surface of the metal slightly to prepare it for paint.

Chapter 12

Castings

PARTS that are made of molten material poured into a mold to solidify are called castings. Such parts are slightly different from the plastic parts in most automobile model kits, for example, because that plastic is injected into the mold under pressure. There are several ways to produce intricate detail castings for a model building project that do not involve the use of molten metal or, for that matter, heat of any kind.

The question is, why do you need or want castings? The casting process is one that allows you to make an almost infinite number of parts using just one carefully detailed original or "master" for the mold. The same principles apply to the vacuum-formed models described in Chapter 4 and the fiberglass hulls shaped over a "buck" in Chapter 10. Casting, however, is a process that is used when very intricate, three-dimensional details are needed in quantity. If you search the hobby shops and hobby catalogs, you may be able to find ready-made castings for the small parts you require. Metal and plastic castings are available for just about every detail part imaginable from scale-size pipe fittings to crisp windows and doors to complex air compressors for steam locomotive models. These detail parts are produced under very carefully controlled conditions from patterns made by the best craftsmen in the world; you cannot hope to match their level of perfection with your own castings. There may be some items, however, that you *can* make by casting to save yourself hundreds of tedious hours making duplicates.

Casting Materials

Most of the castings you can buy are made of one of three materials; plastic, brass, or tin–lead alloys. The plastic parts are produced in expensive injection-molding machines. The brass parts are made by the "lost wax" process. The original pattern for the brass parts was carved and reproduced as an intricate wax casting. Some of the originals are even carved in the wax that is available at lapidary hobby shops or wholesale jewelery supply firms.

The wax pattern is fitted with a sprue of about ⅛-inch diameter wax attached to a flat side of the part. The sprue is suspended above a container about the size of a paper drinking cup so that the part hangs down inside the container. A special plaster is then poured into the cup to completely submerge the part. When the plaster hardens, the cup and plaster are placed inside an extremely hot (about 2000°F) oven or kiln to heat the wax to the point where it literally vaporizes. The cavity left by the wax is then filled with molten brass. When the brass solidifies, the plaster is chipped away and the remaining part is a perfect replica of whatever was carved in the original wax. The sprue is then cut off with a grinding wheel or cutoff disk and the result is a lost wax casting in brass.

This is the identical system that jewelers use to make rings and similar objects in gold or silver. If you carve a wax pattern, a jeweler or lapidary hobby shop can duplicate the pattern for you in gold or silver, but they probably

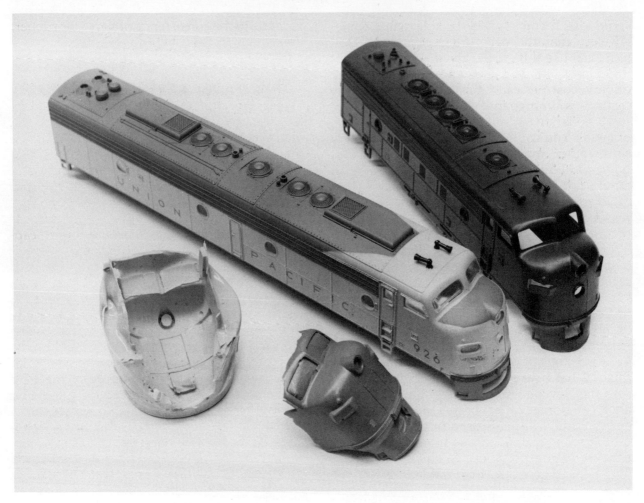

Fig. 12-1 An RTV rubber mold of the F7 diesel's nose *(right)* has been made and a casting of metal-filled epoxy *(foreground)* will replace the nose of the E8 (Union Pacific) diesel.

won't be able to do it in brass because brass has a higher melting point. If you want several duplicates, you'll need to make a mold of the original pattern in the silicone bathtub sealing rubber sold by hardware stores. That mold can be used to make dozens or hundreds of wax replicas of your part which, in turn, can be used to make lost wax castings. Again, a jeweler or lapidary hobby shop can make both the mold and the wax castings. Neither injection molding of plastic or lost wax casting is really worth the trouble or expense for most one-of model projects.

Tin–lead castings do present some possibilities for the modeler. Tin–lead alloy parts purchased ready-made can range anywhere from a single window to a complete 1/43 or 1/24 scale "soft metal" car kit like those mentioned in Chapter 4. These parts are molded in centrifugal casting machines like those sold by Romanoff Rubber Co. The molds are produced by vulcanizing a special rubber disk over an original metal pattern or "master." The disks split in half like the layers of a cake to allow the "master" and the later castings to be removed. A 2-inch-diameter hole is cut in the middle of the top disk of the mold. Small triangular gates or sprues are then connected from the parts around the mold to the central hole. The mold is placed in a machine that spins it with the center hole as the axis. A ceramic funnel is connected to the center hole and molten alloys of tin and lead are poured into the hole. Centrifugal force carries the metal into the cavity to fill it. If you really want to make a number of castings, you can buy a small version of the casting machine, the mold vulcanizer, and the metal's melting pot for about $1000. You may also be able to find some jewelery manufacturers in most major cities (who can use this same type of equipment) to make parts for you at about a tenth of this original cost per each master. The completed castings will range in cost from about a dollar to 10 dollars each, depending on the weight of metal. You could make a similar mold from some other material and simply pour the tin–lead alloy into the cavity, but the metal would not have flowed to many areas of the mold under simple gravity pressure; for this reason the rubber molds are spun to use centrifugal force to help fill the cavity completely.

The most suitable material for making castings in your home workshop is the two-part resin and catalyst similar to that found in conventional epoxies. If you are only going to make a few castings, several tubes of the Duro brand aluminum-filled epoxy is a most suitable material. If you need to produce several parts or larger ones, you may want to locate an industrial hardware or machine tool firm that sells the Devcon brand Devcon C or its equivalent in some other brand of metal-filled epoxy. This material is used to test molds and other tooling in industry so it is quite capable of reproducing the finest details in the mold. The material gives off very little heat as it cures so you do not have to use vulcanized rubber molds as you would with tin–lead alloys; simple room temperature vulcanizing (RTV) rubber molds will do nicely. The epoxy and the RTV do give off toxic and flammable gases, however, so you must work in a well-ventilated area.

The RTV is very similar to the white or black silicone bathtub caulking compound that is sold in tubes at hardware stores. The material is much less expensive if you buy it in pint, quart, or gallon cans. Electronic parts distributors or industrial hardware firms may carry Devcon's Flexane 80 or General Electric's Silastic RTV-31; both work quite well. The master pattern can be carved from jeweler's wax, plastic, wood, or metal or you can duplicate a part from a kit. The part should be coated lightly with a mold release agent like spray-on silcone or Teflon powder. You *must* check both the mold material and the release agent to see if they impair the chemical reaction that hardens the RTV rubber; coat a portion of a scrap of the master's material (coated with mold release) with a layer of the RTV you have chosen and let the compound cure for 48 hours. If it hardens to the consistency of very soft rubber, then the materials should work fine.

Making Molds

The soft rubber RTV molds are not practical for mass producing parts to be sold to the public. The soft rubber will begin to tear and chunk after you have removed one to four dozen castings from the mold. Manufacturers need molds that produce at least several thousand parts to make the process economically sound. The soft RTV rubber gives you an important advantage over hard-surfaced molds because you do not need to worry much about "captures" or the slanted edges of parts (called "relief angles") that are needed to get hard parts to release from metal molds. You must, however, avoid any major captures (Fig. 12-2, *top*) in the mold by locating the angle of the master pattern and the parting line between the two halves of the mold so the mold can be separated easily and the part removed (Fig. 12-2, *bottom*). The flexible nature of the RTV rubber will allow you to spread the mold a bit to clear rivets and other small details.

If there are larger protrusions or captures, those parts must be removed from the master pattern and produced as separate molds and castings. This is the reason that plastic kits, for example, have so many pieces; if the manufacturers could get all those details in a single piece they certainly would. Even then, the metal molds for plastic parts have built-in "ejection pins" to literally drive the parts out of the molds.

The *only* instance in which I would suggest the use of a hard mold would be to produce RTV rubber tires; here, the mold can be machined on a lathe into a pair of aluminum blocks. The flexible nature of the RTV tires would allow you to bend them, if necessary, to get them out of the mold. Don't try molding hard parts in hard molds. Even RTV tires or other soft parts can be molded in RTV molds!!

The RTV compound is soft enough to distort during the molding process. When you make the mold, therefore, use a sturdy container to prevent this. When you have poured the casting material into the finished mold to produce a part, place the mold back into its original sturdy container until the casting is hardened. If the sturdy container is no more than ½ inch larger than any area of the master pattern (and no thinner than about ⅛ inch), quite precise castings can be made using the soft rubber.

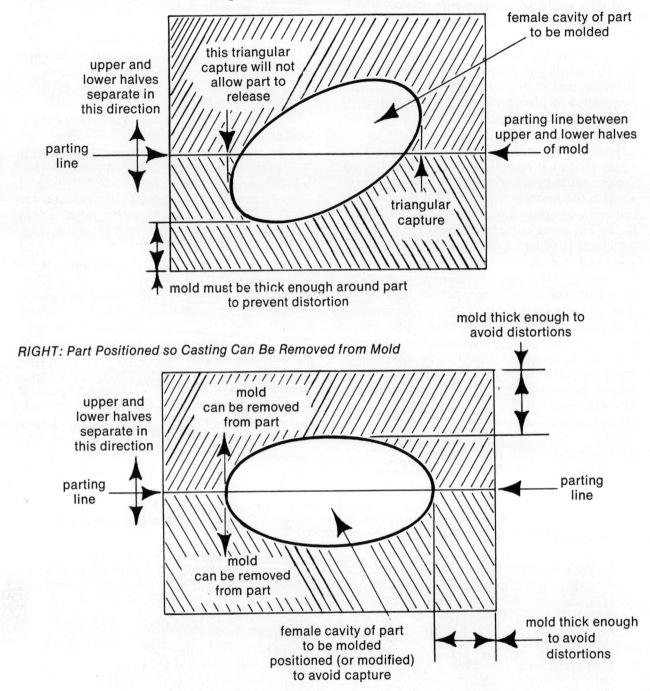

Fig. 12-2 Casting molds: The part in the top two-piece mold is "captured"; the part in the bottom mold is positioned properly.

The container is what keeps the soft rubber from distorting, so it is really an essential "third" part of any RTV rubber mold.

Try to avoid the need for two-piece molds like those shown in Fig. 12-2. It is very difficult, indeed, to force resin or any other casting compound into the cavity of such a mold. Just providing the hole or gate into which you will pour the casting material is very difficult. If I were to make a mold for the egg-shaped part in Fig. 12-2, I would lay the part on its side, as shown in the lower view of the drawings, and submerge all but the center half of the curved side in a single mold. That side could be shaped by grinding and filing the finished part. If a really accurate part is needed on all sides, I would try to make the *casting* in two pieces using separate RTV rubber molds to make (*a*) the top half and (*b*) the bottom half of the part. If you fill the mold full enough to overflow just slightly, that excess material can be ground or filed away so the two halves of the casting can be joined to duplicate the master pattern precisely.

The casting material will generally shrink somewhat as it hardens; therefore, a "shrink factor" must be *added* to all the major dimensions of the master pattern. Each brand of casting resin or epoxy has its own specific shrink factor but most will shrink between .001 and .005 inch per inch. That little shrinkage probably will not matter unless you are casting something really long like the side of a railroad passenger car; if the master pattern for the side was 12 inches long, then the epoxy castings would be 12 inches minus about .060 inches (12 × .005) or 11.040 inches. That .060-inch error could make it difficult to fit the sides to the roof you intend to use. In this case, the master pattern should have been made about 12.060 inches long to produce parts that were truly 12 inches long. The process is really not quite that accurate; you may find as much as .010 inch difference between two epoxy castings made from the *same* mold. You can minimize such differences by being certain to support the mold in the container you used to make it, by precisely measuring and mixing the epoxy resin and catalyst, and by maintaining a constant temperature in the room until all the castings are poured and hardened.

One-Piece Molds

It takes a special type of cleverness to design castings for one-off miniatures because you do not have to worry about the kinds of things that a kit manufacturer must consider. Murray Sincoff is one modeler who has trained himself to think about what he needs in a casting. He wanted to alter the shape on the nose of several E-8 diesels (the Union Pacific body in Fig. 12-1) to more closely match the shape and number board positions of the nose on an F-7 diesel (the black body in Fig. 12-1). Normally, such a complex part would require at least a two-piece mold, but Murray noticed that only the nose itself was needed—no windows or side detail was required. He carefully sealed the window openings and headlight holes with scraps of plastic cemented permanently inside one of the F7 bodies so he could use it rather than having to carve one himself. The body was then suspended above a paper drinking cup with hatpins driven into the sides and roof of the body. The RTV rubber was poured around the body until it filled the

Fig. 12-3 Dale King carved this 1/43 scale Porsche *(left)* from plastic; made an RTV rubber mold of the body *(center)*, and produced cast replicas of the body in clear casting resin (from a craft store) to make thin-skin models *(far right)*.

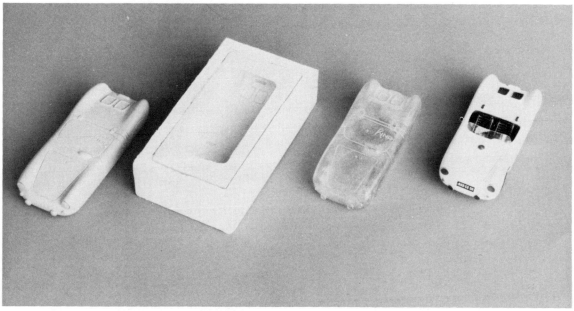

BUILDING MATERIALS AND TECHNIQUES

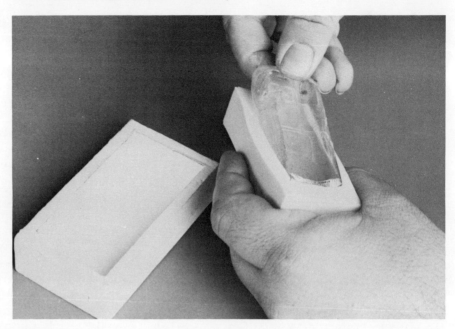

Fig. 12-4 A cast plaster box (*left*) holds the soft RTV rubber mold rigid while the casting resin is hardening. When the resin is hardened, the mold can be flexed to remove the cast body.

cup enough to cover all of the nose. When the RTV hardened, the cup was removed (Fig. 12-5) and Murray had his new nose mold.

Casting with Epoxy

The epoxy and resin mixture will have small air bubbles trapped inside the fluid that will float to the surface just before the mixture cures or hardens in the mold. When you pour the mixture into the mold, additional air bubbles will be trapped between the epoxy and the mold. Those air bubbles can create tiny craters in the surface of the part. Some of those craters can be filled with a dab of epoxy and sanded flush on the finished part. It's best, though, to try to eliminate the air bubbles. Use an old brush to slowly brush about 1/16 inch of the epoxy onto all the surfaces of the mold; this helps to avoid the air bubbles in the surface. The rest of the fluid can then be poured into the mold. It is possible to make a "hollow" casting, like the noses for Murray Sincoff's diesels, by brushing on a layer of resin and letting it cure and then brushing on another 1/16-inch-thick layer and letting it harden. If the resin wants to flow away from the sides (by gravity's force) and into the bottom of the mold, you may need to add a layer of fiberglass cloth to help keep the resin on the sides.

Alternatively, you may need to lay the mold at an angle so you can brush in the epoxy on the horizontal half and, when that cures, tilt the mold so you can brush in the second half on more-or-less horizontal surfaces. You may need to position the mold three or four times to produce some types of thin-wall castings with one-piece molds.

You can try a third method: casting an "interior" plug in the RTV rubber so you have a two-piece mold that will shape both the inside and the outside of the part. The interior plug mold for the diesel noses, for example, would be made by pouring RTV *inside* the body. That

Fig. 12-5 Murray Sincoff used a paper cup as the container for the RTV rubber mold of the F7 diesel nose.

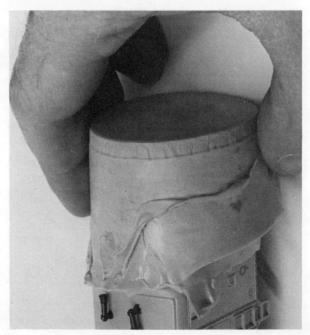

CASTINGS

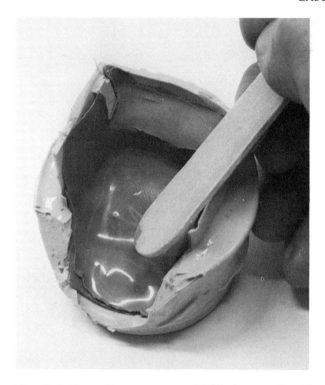

Fig. 12-6 The casting resin or metal-filled epoxy should be brushed over the surface of the inside of the mold. Build up the thickness of the epoxy resin by adding more material with a wooden spatula.

second mold would then be carefully positioned over the first mold and the epoxy would be poured between the two. Air bubbles would be very difficult to avoid even with this simplified version of a two-piece mold.

Using Other Casting Materials

For some models, you may want to substitute plaster of paris for the epoxy as the material for your castings. It is extremely difficult to make large, flat surfaces in resin because such shapes tend to warp as they harden. Magnuson Models makes N, HO, and O scale cast resin structure kits that have very complex patterns for the walls. I suggest you build at least one of their kits before attempting to make a building with cast resin walls using you own homemade molds. Plaster of paris is a much easier material to work with as long as you make the finer details (like window frames) add-on parts that can be either commercial castings or fabricated from plastic or wood strips. Again, I suggest you assemble a structure kit or two that utilizes plaster walls to understand how the system works; Kurton Products, Model Masterpieces, A.I.M. Prod-

Fig. 12-7 The few tiny pits or blemishes in the metal-filled epoxy (Devcon C) casting (*right*) are caused by trapped air bubbles.

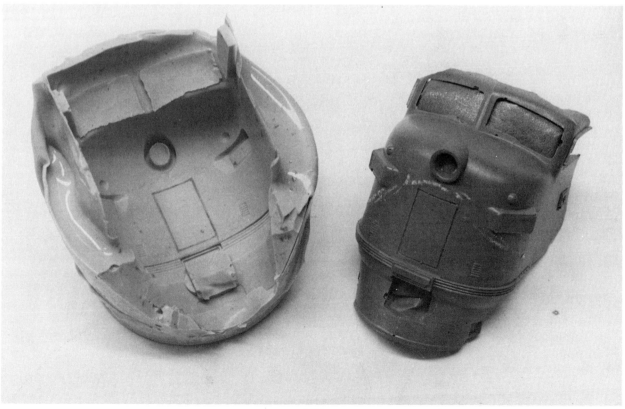

BUILDING MATERIALS AND TECHNIQUES

Fig. 12-8 The walls of this 1/35 scale Kurton Products house kit (for armored model dioramas) are cast in plaster. The windows, however, are separate pieces.

ucts, and Thomas A. Yorke are a few firms that offer cast plaster walls, tunnel portals, and complete structure kits in cast plaster. Yorke even duplicates splintered boards in the material.

The expandable polyurethane foam that is sometimes used to spray on insulation in homes and refrigeration trucks is another casting material that is worthy of consideration for some castings. I.S.L.E.'s Mountains in Minutes series of rock castings and scenery kits includes a two-part polyurethane foam liquid and molds made from latex rubber. The polyurethane expands and hardens into a casting with a smooth skin that captures detail well with an interior structure like a hardened fine-pore sponge. The material's primary advantage is that it is very light weight, which can be helpful for dioramas or model railroad layout modules that must be portable.

If the exact shape of the casting is not too important (for a rock casting to be used as scenery, for example), then the mold can be made from liquid latex rubber. The rubber is available from Mountains in Minutes, and craft supply stores may sell the Berstead brand. Spray the master pattern with silicone lubricant from an aerosol can to serve as a mold release and brush on the liquid latex rubber. The latex tears fairly easily, so you may want to reinforce the mold with a layer of surgical gauze, followed by a second layer of liquid latex. Three or more layers of the liquid latex (with or without the gauze) will be needed for most molds.

Casting Inserts

It is possible to cast small pieces of metal into the parts you are making to provide, for example, a built-in nut to make it easier to attach the part to the complete model or vise versa. These metal inserts must be included in the master pattern's surface so there will be room for them in the mold. The easiest type of fastener to cast into a part is a simple screw extending out from the model. When you are ready to make a mold of the master pattern, drill and tap a hole to accept, for example, a 2-56 screw. Thread about ¼ inch of a ½-inch 2-56 screw into the hole and then cut off the

Fig. 12-9 Mountains in Minutes produces both the liquid latex used to make a mold from a real rock and the expanded urethane foam that was used to make the ½-ounce cast replica of the rock.

screw head with diagonal cutters or a saw. The mold will duplicate that thread so you can simply push a ½-inch 2-56 screw into the hole with the head of the screw hanging out into the mold. When you pour in the casting resin, the fluid will flow around the head of the screw to encapsulate the screw in the casting. When you remove the hardened part from the mold, that 2-56 screw will be protruding from the surface just as it was on the master pattern. You'll need a 2-56 screw for every part you cast, of course.

The exposed screw threads will eventually tear the rubber mold, but you should be able to get at least a half-dozen castings before that happens. You may be able to cover up the tear to plug the hole if you need more castings but they won't have that built-in screw. Devcon C and most of the other metal-filled casting resins or epoxies are strong enough to be drilled and tapped in thread sizes of about 2-56 and larger. The fine threads of 0-80 and smaller screws are too fine for the material.

If you need to have an internal thread that will be used frequently, it's best to cast a nut right into the parts by drilling a hole in the master the size that will *clear* the threads. The mold will then have a little rubber pin that you can use to "hang" a nut on just before you pour in the casting resin. The nut will be captured in the surface of the part. Again, a new nut will be needed for each part you cast. This same technique can be used to "cast in" small metal brackets for detail or to use as attaching points.

Assembling Castings

Epoxy resin castings can be assembled using the same material for a really solid joint. They can be attached to plastic, wood, or metal using either 5-minute epoxy, thickened cyanoacrylate (like Super Jet) or more of the casting resin material. The castings can be filed, sawed, or ground with the same tools that are used to work with metal so the joints will fit precisely with adjacent parts. The epoxy castings are much harder than conventional styrene plastics, however, so it's worthwhile to spend extra time making master patterns and molding castings that fit with a minimum of filing.

Chapter 13
Working Models

THE model building hobby offers a choice of simply assembling replicas that can be admired as museum miniatures or building action machines that perform as realistically as they look. There is certainly no reason to limit your modeling efforts to just display pieces while ignoring the action versions. A large number of static aircraft enthusiasts, for example, also have at least one radio-controlled flying aircraft. Model railroaders often build dozens of models, including completely detailed dioramas or modules with scenery, before they find the time or space to build an operating layout.

There is absolutely no reason that you cannot build a flying airplane, a ship that actually sails, or a radio-controlled racing car with as much detail as the same model might have if built strictly for display. However, in some cases, you may not want to include the fragile details on a model that moves; doorhandles, windshield wipers, or rearview mirrors don't stay in place for long on an electrically powered tabletop "slot" car. If this is your first attempt at "action," you'd be wise to learn from a kit or to build a relatively simple model so you will understand the special requirements of models that move.

Model Engineering

I feel it is only fitting that handcrafted (scratch-built) models receive as much admiration as any other type of sculpted art. Artists who capture real-life scenes realistically on canvas are highly admired in our society, but the artist who creates an even more realistic scene in three dimensions is too often considered to be playing with toys. The painter and the sculptor of miniatures both know and appreciate their own art; the problem is one of an ignorant public. That situation is changing as more and more millions join the ranks of model builders. In time, the general public may even come to appreciate action miniatures for the true works of engineering they are. It is just as difficult to capture all the subtle movements of a full-size sailplane with a radio-controlled miniature as it is to build the real thing. There's certainly less labor and less cost in the miniature, but that is merely a function of its size. In truth, the miniature may be a bit more difficult to fly than its prototype because the on-the-ground flier lacks the seat-of-the-pants feedback that the pilot in the aircraft receives. The same is true of radio-controlled race cars and boats as well as the miniature locomotive, whose engineer must try to couple onto a string of cars without a jolt. Model engineering falls into two basic categories: the power that moves the model and the means by which the model is controlled. Both aspects of the action must be considered when the model is still in the planning stages.

Weight is one of the most important considerations of model engineering. It's not really important how much or how little a display model weighs as long as its wheels or display stand can support it. When the model is sup-

BUILDING MATERIALS AND TECHNIQUES

Fig. 13-1 The Peck-Polymers' "Gangobie" flying "Peanut scale" airplane is powered by a twisted rubber band. Photo courtesy of Peck-Polymers.

posed to move, though, weight is an extremely important factor. In general, the body (or whatever the visible superstructure is called) should be made as light as possible. This will allow you to place any additional weight exactly where it can do the most to help the performance of the model. The Revell 1/32 static model plastic aircraft kits can be assembled into miniatures that weigh no more than some similarly sized models from Cox with gasoline engines for power and tether or "U-control." The weight of the display model is all in the thickness of its plastic while the flying model has a stronger plastic that is made very thin to leave enough "room" for the engine and fuel tanks and, perhaps, some lead to balance the model fore and aft.

You will find that it is almost impossible to build a radio-controlled aircraft or race car that is too light; it's difficult to make the bodies and

Fig. 13-2 This 1/32 scale Airfix brand injection-molded plastic Bentley has been modified so it can be raced on a slot car track.

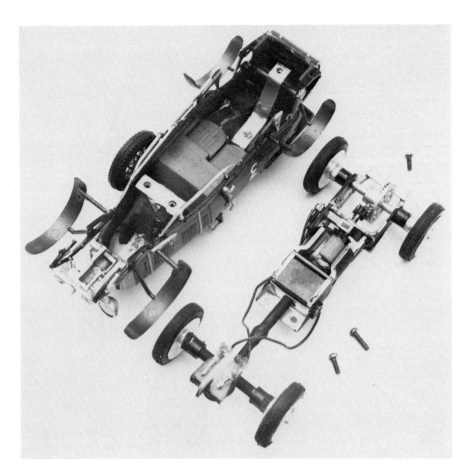

Fig. 13-3 K & S brass tube and brass strip were soldered together to make the chassis for the 1/32 scale Bentley. The mounting brackets were drilled and tapped (threaded) to accept 2-56 screws and then attached to the body with epoxy.

Fig. 13-4 The ready-to-run and kit-built chassis, like this Mantua/Tyco 1/87 scale 2-8-2, are the easiest shortcuts to building a powered scratch-built model. Simply build a new superstructure and tender body to match some particular prototype plan.

frames of such models strong enough without adding so much weight to the design that their performance is poor. Conversely, it is almost always necessary to add weight to models of steam or diesel locomotives, to ships' hulls, and to tanks and other armored vehicles. Trains and tanks need the weight to help give them more traction or pulling power; ships need the weight to keep from tipping over in a turn or in high winds. It takes years of experience to know which models demand how much weight, but that experience is easy (and fun) to gain because it can come from simply building a variety of action models from kits.

Scale Power

The source of energy for your model can be as simple as a rubber band or as complex as a precise scale replica of the engine that powered the prototype. Modelers in this electronic age all too often forget that a miniature can be powered by a simple rubber band engine. The rubber bands that power Peanut scale model airplanes like those described in Chapter 3 will also work in models of all types of power boats. Don't overlook the simple wind-up or clockwork motors either; the clockwork mechanism was popular in the twenties and is still available in all types of toys. It's easy enough to adapt a proper size wind-up motor or the complete chassis to power a scale model boat, tank, automobile, or even a railroad locomotive. Using an internal combustion engine to power a model of a machine that had a similar kind of motor is a most satisfying aspect of model building. You'll find two-stroke engines with mounting brackets or even complete chassis to fit fuel-burning engines in all types of aircraft, in boats, and in 1/12 or 1/8 scale automobiles.

The electric motor, especially one powered by on-board batteries, is certainly the engine

Fig. 13-5 "Toy" radio-controlled boats, like this Pro-Cision "Sun Cruiser," can supply inexpensive power and control systems for similar-size scratch-built boats.

of our era. Dozens of plastic model kits include inexpensive motors, gears, and battery boxes to power boats, tanks, and automobiles. The rechargeable nicad batteries make it worthwhile to consider the use of electric motors for any type of model, including such highly sophisticated aircraft as helicopters. Again, the best way to learn how to use electric motors for power is to try them in a few ready-made chassis so you can understand how to design your own.

Control Systems

The engineering effort required to make a miniature perform is most visible when you consider the control systems that are now available for all types of miniatures. The fundamental system of control is what aircraft modelers call "free flight"; when the model is launched it is controlled only by the built-in settings of the rudder and other control surfaces. Battery-powered boats, tanks, and automobiles are equivalents of free flight; their steering systems are set for either straight ahead or a curve and the motors go full blast until they run out of energy. Wind-up or clockwork toy trains operate on a similar principle except that they have the tracks to guide them until their motors run down.

Tethered Control

The simplest of all the remote-control systems falls into the category of tethered control. The tether is a string or wire or even an electrical cable that connects the model to the operator (or pilot, driver, or engineer). Some of the tethered-control systems are purely mechanical devices where the operator moves a lever in his or her hand-held controller. That lever pulls one of the tethers or cables to move a similar lever inside the model. This is the system used (with two cables) to control the elevator for up and down movement of a U-control flying model airplane. A number of toys also have cable-control systems like this.

The somewhat more modern systems of tethered control use electrical signals to turn magnetic relays or motors on or off on the

Fig. 13-6 This "Group 12" radio-controlled racer, from BoLink, is a standard size for 1/12 scale electric-powered radio-controlled racing cars—you can supply a scratch-built vacuum-formed body.

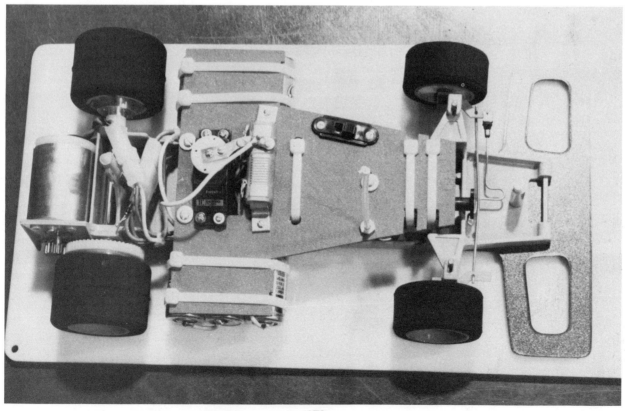

BUILDING MATERIALS AND TECHNIQUES

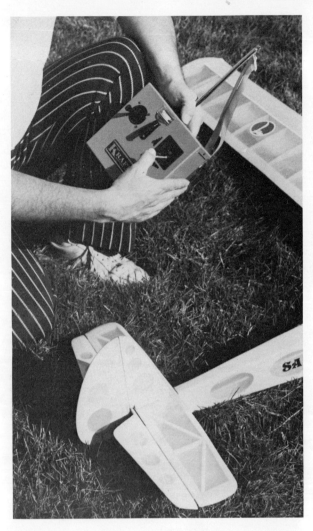

Fig. 13-7 This Kraft radio-controlled transmitter has three channels to control the rudder, elevator, and (on this sailplane) the flaps on the wings for banking control.

model. Many of the toy robots and boats have this type of system to control direction and to turn the motor on or off. Almost all model railroads (and the electric slot racing cars) use a type of tether system in which the two (or three) rails of the track carry the power as though they were electrical cables. The rails serve a double purpose, with model railroads, in that they also serve to guide the trains' direction. The slot guides the slot car around the track.

The advantage of the tethered system for control is that the power source can also be remote, so the model does not have to carry bulky batteries. There are a few electric airplanes on the market whose motors are controlled by tethered wires running to a control box in which the batteries to power the motor are contained.

Radio Control

The concept of sending control commands through the air to the model is undoubtedly the most exciting development in the hobby of miniature vehicles. The concept of radio control is easy to explain. The modeler holds a transmitter in his or her hands. This transmitter sends out radio signals similar to those sent out by a television or radio station's transmitter. The receiver in the model is just a variation of a television set or radio except that the transmitter's signals do not control sound or pictures. The radio-control receiver in the model sends more or less electrical power to one or more tiny motors in the model. The transmitter, then, controls motors (called "servo motors") in the model through the receiver in the model. The servo motors are linked to levers to move, for example, the elevator and rudder to control an airplane's movements up and down or to the right or left. Several types of speed controllers are used to allow radio control of the electric motors used to power the propellers in aircraft and boats or the wheels in cars or tanks. A servo motor and levers move the throttle controls on the carburetor in models equipped with fuel-burning engines.

The radio-control systems in toys don't usually have servo motors; the radio signals simply turn switches on or off so that you have either full power or none, full right or full left turns, and so on. The more expensive toys and the radio-control systems sold for models have what is called "fully proportional" control so that the movement you make on the lever in the transmitter produces exactly the same degree of movement in the model. The very precise steering and throttle control needed for a race car or airplane or sailboat is achieved with this type of system.

Each servo motor or on-off switch in a radio-control rig is called a "channel." A separate channel is needed for each control, so an airplane might have a channel for the elevator (to control up and down movements) and a second channel for the rudder (to control right and left movements). Some of the "beginner" systems have a single channel that, in the case of the airplane, controls the rudder. The motor would run long enough to pull the airplane to a hundred-foot altitude and the wings and

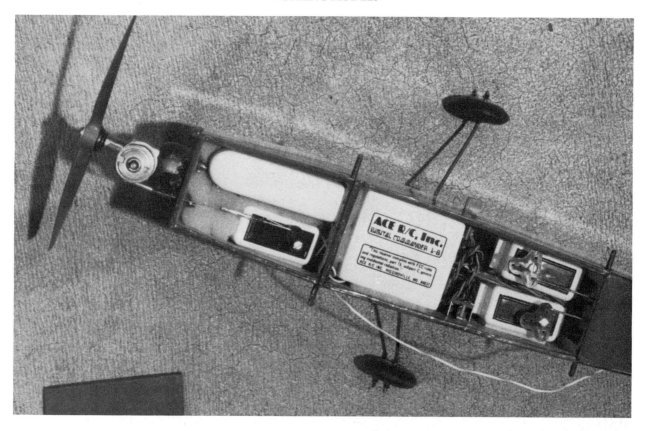

Fig. 13-8 The Ace R/C radio-control receiver with three servo motors is one of the new compact designs. Here, the third servo motor (channel) is being used to control the engine's speed.

elevator would be set so the plane would glide. The only really necessary control would be the rudder. The single channel would control the steering system appropriate to a boat, car, or tank. Two channels are enough to control racing cars, boats, or tanks—one for speed and the other for direction. The braking action on a car or tank is simply a bit beyond "off" on the lever that controls the motor or engine speed. Aircraft models may need three channels if you want to control the speed of the model.

The cost of the radio-control outfit is almost directly related to how precisely it controls each servo motor, how many channels or servo motors it has, and how powerful the transmitter and receiver may be. The power is really only important if you are flying aircraft; even the better toy sets have enough power for automobiles, tanks, and boats. A "toy" can supply simple on-off controls for two channels and cost less than $50. Two-channel, fully proportional systems start at about $100 with quick-response systems selling for about twice that. The systems range upward to over $500 for a seven-channel rig.

Radio Installations

The radio-control system will add quite a bit of weight to your model because the receiver and servo motors must have on-board batteries for power. There are a variety of sizes of receivers and servo motors on the market, so you can match a small radio-control rig to a small model. Larger models will require larger and more powerful servo motors to operate the controls. For the moment, the state-of-the-art of radio-control components limits the minimum size of an aircraft to be powered by radio control to one with about a 24-inch wingspan.

It's difficult to fit existing radio-control receivers and servo motors into boats, automobiles, or tanks any smaller than 1/20 scale. There are, however, some 1/24 scale radio-controlled "toy" racing cars that have their electric motors, servo motors, and receivers in one compact integrated circuit unit. You could fit an alternative body to one of these toys. The weight of your model may be increased considerably if you decide to use an electric motor to power it because that motor must also have its own on-board batteries. You can save some weight (and bulk) by using the motor's batteries to power the radio-control system on a racing car. That's not wise with an airplane or boat because the model may go out of your control when the motor drains the batteries' power; those models should have separate battery packs for the motor and for the radio rig.

I recommend that you learn to fly (or drive or sail) by radio control with one of the better radio-controlled toys. You'll know the radio installation is right with this type of model, and you can probably crash the model once or twice without destroying it as you might a more delicate scale model. It is seldom wise, however, to attempt to transfer the radio receiver and servos from a toy aircraft or car to a hand-built model. Purchase one of the two or three-channel radio-control transmitter, receiver, and servo motor rigs from a hobby dealer. His advice will be invaluable in helping to match the requirements of your proposed radio-controlled model to those servo motors, receiver, and transmitter.

Sources of Supply

THERE are several thousand firms that manufacture products for the model builder. A listing of all of them would fill a volume larger than this one. If you are searching for a particular firm, I suggest you check with your local hobby dealers first and, if they cannot help, search through the ads in the hobby magazines and the sources of supply lists in hobby books.

The firms listed here can usually supply a catalog, but there is often a charge. Some of the catalogs are a simple one-page listing while others are giants like Walthers' 400-plus-page HO railroad catalog. When you write for catalog information, be specific about what items or categories of items you are interested in. Your specific hobby interest may be only a small part of that firm's stock of products.

If you do write to any of these firms for the price of a catalog or with a question, you must enclose a stamped, self-addressed envelope if you expect a reply. You might also mention where you heard of the existence of the firm. Some of these companies are part-time businesses operated by hobbyists; others are merely divisions or departments of much larger corporations. In either case, it may take several months for a reply to reach you. You can expect much better success if you can persuade your local hobby dealer to order whatever catalogs you need or to ask your questions for you. You can locate hobby shops in your area by looking in the Yellow Pages under the heading "Hobby & Model Construction Supplies—Retail."

Model Plans

Plans for Rockets

Ballantine Books
 201 E. 50th St.
 New York, NY 10022

Centuri Model Rocketry
 1095 E. Indian School Rd.
 Phoenix, AZ 85014

Estes Industries
 P.O. Box 227
 Penrose, CO 81240

Gamescience
 01956 Pass Rd.
 Gulfport, MS 39501

Model & Allied Publications
 P.O. Box 35, Bridge St.
 Hemel Hempstead, Herts HP1 1EE
 England

StarFleet Division
 5035 Swingle Dr.
 Davis, CA 95616

Plans for Aircraft

Aeroplane Monthly
 IPC Transport Press Ltd.
 Dorset House, Stamford St.
 London SE1 9LU
 England

AirInternational
 Fine Scroll Limited (England).
 P.O. Box 353
 Whitestone, NY 11357

D.M. Emmons
 P.O. Box 122
 Batavia, NY 14020
 (paper models)

Flying Models
 Box 700
 Newton, NJ 07860

John Hathaway
 P.O. Box 1287
 San Pedro, CA 90731
 (paper models)

International Plastic Modelers
Society
P.O. Box 2555
Long Beach, CA 90801

Model Airplane News
837 Post Rd.
Darien, CT 06820

Model & Allied Publications
(see Plans for Rockets)

Model Aviation
815 15th St. N.W.
Washington, DC 20005

Model Builder
621 W. 19th St.
Costa Mesa, CA 92627

Modernistic Models
Box 6974
Albuquerque, NM 87197

Sid Morgan Vintage Aviation
Plans
13157 Ormond
Bellville, MI 48111

Peck Polymers
P.O. Box 2498
La Mesa, CA 92041

John Pond Plans
Box 3215
San Jose, CA 95156

Radio Control Modeler
Box 487
Sierra Madre, CA 91024

Scale Models
P.O. Box 35, Bridge St.
Hemel Hempstead, Herts HP1 1EE
England

Plans for Cars

Bob Clidinst
5771 W. 44th St.
Indianapolis, IN 46254
(1/24-scale racing auto plans)

James Ison
2509 E. Jackson Blvd.
Elkhart, IN 46514

Model & Allied Publications
(see Plans for Rockets)

Road & Track
1499 Newport Ave.
Newport Beach, CA 92663

Scale Auto Enthusiast
P.O. Box 441
Menomonee Falls, WI 53051

Scale Models (magazine)
(see Plans for Aircraft)

Plans for Ships

James Bliss Marine
Dedham, MA 02026

Floating Dry Dock
c/o General Delivery
Kresgeville, PA 18333

John Hathaway
(see Plans for Aircraft)

Model Boats
P.O. Box 35, Bridge St.
Hemel Hempstead, Herts HP1 1EE
England

Model Ship Builder
P.O. Box 441
Menomonee Falls, WI 53051

Model Shipways Co., Inc.
39 W. Fort Lee Rd., Box 85
Bogota, NJ 07603

Model Shipwright
Conway Maritime Press Ltd.
2 Nelson Road
Greenwich, London SE10
England

The National Maritime Museum
Greenwich
London SE10
England

Nautical Research Guild, Inc.
6413 Dahlonega Rd., Md.
Washington, DC 20016

Smithsonian Institution
c/o Curator of Transportation
Washington, DC 20560

Taubman Plans Service
11 College Dr., Box 4G
Jersey City, NJ 07305

Edward H. Wiswesser
407 N. 25th St.
Pennside, Reading, PA 19606

Plans for Armored Fighting Vehicles

International Plastic Modelers
Society
(see Plans for Aircraft)

Military Models
P.O. Box 35, Bridge St.
Hemel Hempstead, Herts HP1 1EE
England

Model & Allied Publications
(see Plans for Rockets)

Plans for Trains

Newton K. Gregg Publisher
P.O. Box 1459
Rohnert Park, CA 94928

Kalmbach Publishing Co.
1027 N. Seventh St.
Milwaukee, WI 53233

Live Steam
P.O. Box 581
Traverse City, MI 49684

Mainline Modeler
P.O. Box 5056
Lynnwood, WA 98036

Model Railroader
1027 N. Seventh St.
Milwaukee, WI 53233

Model Railroading
2901 Blake St.
Denver, CO 80205

Narrow Gauge and Short Line Gazette
P.O. Box 26
Los Altos, CA 94022

Simmons-Boardman Publishing Corp.
508 Birch St.
Bristol, CT 06010

Plans for Structures

American Life Foundation
Watkins Glen, NY 14891

Builder Plus
Polks Hobby Dept. Store
314 Fifth Ave.
New York, NY 10001
(paper models)

Dover Publications, Inc.
180 Varick St.
New York, NY 10014

Newton K. Gregg Publisher
(see Plans for Trains)

John Hathaway
(see Plans for Aircraft)

Kemtron Sales
P.O. Box 214235
Sacramento, CA 95821

101 Productions
834 Mission St.
San Francisco, CA 94103
(paper models)
Note: the model railroad magazines include structure plans and articles in most issues.

SOURCES OF SUPPLY

Building Materials

Arbour Models
P.O. Box 1352
Syracuse, NY 13201
(castings for trains)

Cal-Scale
P.O. Box 475
Pinedale, CA 93650
(brass train castings)

Caltronic Laboratory
P.O. Box 36356
Los Angeles, CA 90036
(metal rods, sheets, and screws)

Craftsman Specialty Supply
6608 Forty Mile Point
Rogers City, MI 49779
(metal, wood, plastic, and castings)

Craftsman Wood Service Co.
1735 W. Cortland Ct.
Addison, IL 60101
(special wood sheets and blocks)

Devcon Corporation
Danvers, MA 01923
(casting epoxy and RTV rubber)

Dromedary Ship Modelers Associates
6324 Belton Rd.
El Paso, TX 79912
(parts for ship models)

Evergreen Scale Models
2685 151st Place, N.E.
Redmond, WA 98052
(sheet and strip styrene plastic)

A. J. Fisher, Inc.
1002 Etowah St.
Royal Oak, MI 48067
(parts for ship modelers)

Grandt Line Products
1040B Shary Ct.
Concord, CA 94518
(plastic train and structure castings)

Holgate & Reynolds
601 Davis St.
Evanston, IL 60201
(brick and stone embossed sheets)

I.S.L.E. Laboratories, Inc.
10009 E. Toledo Rd.
Blissfield, MI 49228
(polyurethane foam and latex casting materials)

K & S Engineering
6917 W. 59th St.
Chicago, IL 60638
(metal sheets, tubes, rods, and wire)

Kemtron Sales
1120-A Gum Ave.
Woodland, CA 95695
(castings and parts for trains)

Locomotive Workshop
Box 211 B 1, RFD 3
Old Bridge-Robertsville Rd.
Englishtown, NJ 07726
(etched brass parts and kits for trains)

Harold F. Mellor
Box 509
Parker, CO 80134
(1/87-scale etched brass train parts)

Midwest Products Co.
400 S. Indiana St.
Hobart, IN 46342
(wood strips and sheets)

Milled Shapes, Inc.
1701 N. 33rd Ave.
Melrose Park, IL 60160
(brass structural shapes and sheets)

Model Expo, Inc.
23 Just Rd.
Fairfield, NJ 07006
(parts for ship models)

Model Parts, Inc.
212 Murray Rd.
Newark, DE 19711
(plastic structural shapes)

Model Shipways Co., Inc.
39 W. Fort Lee Rd., Box 85
Begota, NJ 07603
(parts for ship models)

Northeastern Scale Models, Inc.
Box 425
Methuen, MA 01844
(wood structural shapes, strips, and sheets)

Plastruct, Inc.
1020 S. Wallace Pl.
City of Industry, CA 91748
(plastic structural shapes and sheets)

Precision Investment Associates (P.I.A.)
P.O. Box 57
Edmonds, WA 98020
(castings and parts for trains)

Precision Miniatures
637 Paularino
Costa Mesa, CA 92626
(wheels and parts for cars)

Precision Scale Co.
1120-A Gum Ave.
Woodland, CA 95695
(castings and parts for trains)

Propar Design
P.O. Box 639
Escondido, CA 92025
(parts and winches for boats)

Robert E. Sloan
30 E. Pleasant Lake Rd.
St. Paul, MN 55110
(etched brass parts for trains).

Special Shapes Co.
P.O. Box 487
Romeoville, IL 60441
(brass structural shapes, rods, and sheets)

Wm. K. Walthers, Inc.
5601 W. Florist Ave.
Milwaukee, WI 53218
(parts, materials, and screws)

Wills Finecast
Lower Road
Forest Row, Sussex RH18 5HE
England
(wheels for 1/24-scale cars)

Tools for Modelers

Badger Air Brush Co.
9201 Gage Ave.
Franklin Park, IL 60131
(airbrushes and accessories)

Binks Manufacturing Co.
9201 W. Belmont Ave.
Franklin Park, IL 60131
(airbrushes and accessories)

Brookstone Co.
127 Vose Farm RD.
Peterborough, NH 03458
(special and unique tools)

Caltronic Laboratory
P.O. Box 36356
Los Angeles, CA 90036
(lathe and milling tools)

Contact, Inc.
9 Elm St.
Hudson, NH 03051
(resistance-soldering outfits)

Dremel Div. of Emerson Electric
4915 21st St.
Racine, WI 53406
(motor tools and power tools)

(American)Edestaal, Inc.
　1 Atwood Ave.
　Tenafly, NJ 07670
　　(lathes and milling machines)

Emco-Lux Corp.
　2050 Fairwood Ave.
　Columbus, OH 43207
　　(lathes and milling machines)

The Foredom Electric Co.
　Bethel, CT 06801
　　(high-speed motor tools)

Formicator (see Idea
　Development)

General Tool (see Walthers)

H & R
　c/o The Original Whistle Stop,
　　Inc.
　3745 E. Colorado Blvd.
　Pasadena, CA 95821
　　(resistance-soldering outfits)

Idea Development, Inc.
　P.O. Box 7399
　Newark, DE 19711
　　(vacuum-forming machines)

K & S Engineering
　6917 W. 59th St.
　Chicago, IL 60638
　　(metal-working tools)

Machinex (see Edestaal)

Microflame, Inc.
　3724 Oregon Ave. S.
　Minneapolis, MN 55426
　　(butane-soldering outfits)

NorthWest Short Line
　Box 423
　Seattle, WA 98111
　　(special cutting and forming jigs)

PBL
　Box 749
　Chama, NM 87520)
　　(resistance-soldering outfits)

PFM (Pacific Fast Mail)
　P.O. Box 57
　Edmonds, WA 98020
　　(vernier calipers)

Paasche Airbrush Co.
　1909 Diversey Parkway
　Chicago, IL 60614
　　(airbrushes and accessories)

PanaVise Products, Inc.
　2850 E. 29th St.
　Long Beach, CA 90028
　　(vises and presses)

Precision Manufacturing Co.
　4546 Sinclair Rd.
　San Antonio, TX 78222
　　(spray booths, presses, and punches)

Romanoff Rubber Co., Inc.
　153 W. 27th St.
　New York, NY 10001
　　(centrifugal casting machines)

Sherline Products
　1320-5 Grand Ave.
　San Marcos, CA 92069
　　(lathes and milling machines)

Stay-Brite Solder
　J. W. Harris Co., Inc.
　10930 Deerfield Rd.
　Cincinnati, OH 45242
　　(solders and fluxes)

Thayer & Chandler
　442 N. Wells
　Chicago, IL 60610
　　(airbrushes and accessories)

Unimat (see Emco-Lux)

Wm. K. Walthers, Inc.
　5601 W. Florist Ave.
　Milwaukee, WI 53218
　　(hand and power tools)

X-Acto
　45-35 Van Dam St.
　Long Island City, NY 11101
　　(hand tools and airbrushes)

Clubs and Organizations

Academy of Model Aeronautics
　(AMA)
　815 15th St. N.W.
　Washington, DC 20005
　　(flying model aircraft)

American Model Yacht
　Association (AMYA)
　2709 S. Federal Hwy.
　Delray Beach, FL 33444
　　(powered boats and ships)

International Model Power Boat
　Association (IMPBA)
　24310 Prairie Lane
　Warren, MI 48089

International Plastic Modelers
　Society (IPMS)
　P.O. Box 2555
　Long Beach, CA 90801
　　(static model aircraft, armor, ships, rockets, and cars)

Model Car Collectors Association
　(MCCA)
　14201 Woodwell Terrace
　Silver Spring, MD 20906
　　(plastic display model cars)

NAMBA International
　Route A, Box 19
　Lower Lake, CA 95457
　　(model boats and ships, powered)

National Association of Rocketry
　(NAR)
　P.O. Box 725
　New Providence, NJ 07974
　　(flying model rockets)

National Model Railroad
　Association (NMRA)
　P.O. Box 2186
　Indianapolis, IN 46206
　　(static and operating railroad models and structures)

Radio Operated Racing, Inc.
　(ROAR)
　7822 Eby Lane
　Overland Park, KS 66204
　　(radio-control automobile models)

Index

ABS plastic(s)
  scribing, 129
  sheets, 36, 69, 124, 126, 127
  tubing, 83
  using, 75, 123
  in vacuum-forming, 62
Ace R/C radio-control receiver, 175
Acid
  etching, 131, 150
  flux, 155–156, 158
Acrylic paints, 143
AFV. See Armored fighting vehicle
A.I.M. products, 165–166
Airabonita fighter, 3
Airacobra fighter, 3
Airbrush
  painting with, 13–15, 45–46, 81, 119
  selecting, 11–13
Aircraft
  control. See Control systems
  cowls, 35, 38, 132, 140
  cutout kits, 109
  electric, 174
  elevator, 174–175
  flying, 39–40, 140, 173–174
  flying model kits, 4, 61, 139
  full-size, 9, 40–41
  fuselage, 23, 36, 38–41, 44–45, 136, 140
  jet, 22, 32, 33
  materials for, 91
  models, 1, 19, 32–47, 67, 132, 135
  parts of, 6, 24, 35, 38–39, 41, 140
  plans for, 177–178
  plastic kits, 15, 170
  propeller, 1, 44–45
  rudder, 39, 174–175
  single-seat fighter, 35
  stabilizer, 39, 41, 45, 47, 140
  trainer, 35
  vacuum-formed kits, 35, 61, 132
  wings, 9, 10, 23, 38–39, 44–45, 136, 140, 174
  World War II, 20, 34
Airfix kits, 24, 32, 80, 83, 170
Alignment blocks and tools, 124, 125, 144
Allard J2, 54
Aluminum
  flush-riveted, 6
  paint, 81–82
  shaping, 136
  soldering, 145
Ambulances, military, 80
American Locomotive Works, 106
AMT kits, 32, 48, 50, 54, 83, 85

Armor, 48, 79, 85–91, 127
Armored conversions, 82–83
Armored fighting vehicle (AFV)
  hull, 82–83, 89
  parts of, 83
  plans for, 178
  vacuum-formed kits, 132
Armored personnel carrier, 79, 90
Armored scouting vehicle, 79
Artillery, 79
Assembly table, using, 129
Athearn Freightliner truck kit, 50
Aurora models, 83
Automobile(s)
  body, 7, 50, 54, 60, 62, 129, 136
  building, 106, 135
  carving, 10
  chassis. See Chassis
  classic, 2, 51
  conversion of, 50–54
  metal miniatures, 49, 54
  models, 1, 19, 33–34, 48–63, 67, 159
  parts, 50, 54, 58–60
  plans for, 9, 178
  plastic model kits, 48–49, 54
  R/C, 48, 54, 58, 175–176
  vacuum-formed, 132
  wheels, 36, 50–51
Balsa, 137, 141
  for aircraft, 139–140
  cutting, 27, 126
  fillercoat, 30
  for flying rocket, 25, 27–28
  hard, for flying aircraft, 44, 139
  hard, for ship hull, 74
  ribs, 45
  sheets of, 39, 41
  strips, 125
  for structures, 116
Basswood
  sheets of, 140
  for ship hulls, 74
  strips of, 63, 112, 140, 142
  for structure kits, 113, 116
Battery(ies)
  nicad, 173
  on-board, 172, 175–176
  -powered models, 173–174
"Battle Corvette Nebula," 31
Bell, milling, 151
Belt grinder, 147
Benchwork, 11, 125
Bentley, 170–171
Binks airbrush, 12
Biplane, World War I, 32
Boat(s)
  building, 135
  control of, 175
  power, 172–173
  R/C, 169, 172, 175–176
Bogie suspension, 85, 86
Bondo putty, 134
Brass
  cab, 99–101
  carving, 123
  casting, 159–160
  etched, 96–99, 150–151

kits, 96–99, 101
locomotive, handmade, 106
rivets in, 131
shapes, 145
sheet, 4, 92, 101, 150
soldering, 157–158, 171
tubing, 69, 145, 171
Brick, simulating, 107, 116–117, 123
Bridge(s), 135, 141–143
Bronze, 145
Brookstone Tool Company, 146–147, 156, 158
Brush, paint, 10–11, 81, 119
"Buck," use of, 135–136, 138, 140, 159
Builder Plus kits, 109–110, 116
Bulkhead(s), ship, 69, 72, 74–77
Burago metal auto kit, 49
Bus, model, 49
Butane, 157
Butyrate, 61–62
Cabin, log, 107
Caboose
  kit conversion, 93–96
  model, 92–94
Cal Scale, 99, 106
Caliper, Vernier, 4, 11, 124
Caltronic Labs, 145, 152–153
Camouflage, 80, 83, 90
Campbell materials, 116, 141–142
Carbon electrode system, 144
Cardboard kits, 107, 109–110
Cardboard parts, 11, 15, 24, 107–109, 111–112, 123
Carpenters' square, 124–125
Carr Lane Manufacturing Company, 152, 155
Carving, 10, 136–137
Casting(s)
  assembling, 167–168
  brass, 159–160
  detail, 5, 92–93, 101, 105–106, 135, 142, 159
  hollow, 164
  inserts, 166–167
  machine, 160
  materials, 159–166
  metal, 142, 159
  plastic, 159
  precut, 93
  rock, 166
  sand, 127
  tin-lead, 159–160
  wax, 159–160
Castle, model, 107, 109–110
Cathedral models, 107, 123
Celotex, 44, 142
Cement glues
  bottled, 23–24
  contact, 78
  cyanoacrylate, 24, 49–50
  flammable, 11
  resin-based, 117
  rubber, 81
  tube-type (thick), 23–24
Cement, simulating, 116
Centuri models, 25, 31
Champ decal set, 14
Chassis
  armored fighting vehicle, 82–83

automobile, 48, 50, 54, 58, 62, 171
locomotive, 171
ready-made, 172
Chemical milling, 147–148, 150–151
Chevrolet AMT 1939, 2
Chisel, 59, 136
Clamps, 23–24
Colorado Midland, 93–96, 99
Compressor, air, 12–13, 159
ConCor/Heljan kits, 116
Constitution, U.S.S., 63
Control systems. See also Radio control
  boats, 172
  free flight, 41, 173
  tethered, 173–174
  U-control, 41, 170, 173
Conversions
  aircraft kit, 35
  armored military vehicles, 80, 82–83
  cab kit, 96–100
  caboose kit, 93–96
  car, 50–54
  ship kits, 68
  vacuum-formed, 35, 79
Copper
  plates, for ship hulls, 76, 78
  shapes, 145
  sheeting, 63–64, 78
Corgi metal automobiles, 49
Cork railroad roadbed, 92
Corvette, 1965, 49
Cottage model, 107
Cox plastic aircraft, 32, 170
Craftsman brand tools, 124
Craftsman Specialty Supply, 145
Craftsman kits, 92–93, 117
Crossing Gate Model kit, 93
Cutout models, 109–110
Cutters, 10, 97, 131, 145, 167
Cutty Sark, 63
DDV flat varnish, 78
Decals, 14–15
Detail Associates, 142
Devcon C, 161, 165, 167
Dioramas
  armor model, 79–80, 166
  armored vehicles, 107
  casting, 166
  farmhouse, 141
  railroad, 91, 169
  waterline ships, 107
Dirigible(s), 32
Display model(s). See Static Models
Dockside scenes, 107
Dollhouse(s), 107–108
"Double Trouble" rocket, 25–28, 31
Dremel tools
  belt grinder, 147
  bits, 146–147, 152
  cutoff disk, 155
  motor tool, 69, 136, 138, 146
Drill(s), 11, 152, 166–167
Drill press, 10–11
Dumas models, 65
Dupont Lucite, 123

INDEX

Duratite putty, 134
Duro epoxy and putty, 61, 134
Dyna-Models, 142
E & B Valley kit, 93
Edestaal, 151
Electronic kits, 96
Embossing, 11, 129–131, 150
Emco-Lux, 151
Engine(s)
  aircraft, 140, 172
  automobile, 20, 50, 172
  flying rocket, 25–27
  fuel-burning, 174
  gasoline, 170
  internal combustion, 32, 172
  for rocket, 25
  for ships, 67
Entex M.G. TC, 54
Enzmann Starship, 21, 23–24
Epoxy(ies)
  aluminum-filled, 106, 161
  casting with, 163–165, 167, 171
  5-minute, 24
  paint, 59–60
  resin, 58
  use of, 38
Ersin rosin-core 60/40 solder, 144, 156
Ertl metal automobile kit, 49
ESCI, 80, 82
Estes Industries kits, 25, 27, 31
Etching
  brass, 96–100
  rivet heads, 128
  sheet metal, 147–151
  technique, 11, 144
Evergreen
  iron works kit, 118–119
  kits, 93, 108
  milling bits, 130
  plastic sheets, 4, 93
  plastic strips, 94–95, 117, 123–124
  railroad car sidings, 93–94
  scale models, 36, 92
  structure kits, 117–119
  styrene, 69, 85, 124, 130
Evergreen Hill Designs, 142
Farmhouse, 141
Fence, 91, 129, 135, 140
Ferrari (1964) Formula I, 54–56
Fiberglass
  cloth, 36, 56, 138, 164
  hulls, 65, 135, 159
File(s)
  card, 144
  coarse-cut, 126–127
  fine-cut, 144
  jeweler's, 10
  medium-cut mill, 144
  mill, 10
Filing metal, 147
Filler putty, 133–134
Fillets, making, 133–134
Fine Scale models, 142
Finish(es), 30, 136–138
  fiberglass, 138
  flat, 12, 15, 78
  glossy, 12, 78
  semigloss, 78
Flexane, 80, 161
Flight Systems, 25

Floating Dry Dock Company, 68
Floor, model, 117
Floquil stain, 143
Flux acid, 155–157
Flying saucers, 22
Flying Stick models, 40–45
Forceps, 68
Formicator vacuum-forming machine, 56, 61–62, 132
Foundry, model, 118–119
Franklin Chemical, 44
Free flight, 41, 47, 173
Freight car(s), 92–93, 96, 151
Freightliner truck kit, 50
Fuel cell, 24
Fujimi AFV kits, 80, 83
Gabriel (Hubley), metal automobile kits, 49
"Gangobie" airplane, 170
General Tool
  machinist's square, 124–125
  pin vise, 95, 144
  steel ruler, 124
Generator, steam, 99
German Panzer MkIII, 82, 84
Glass, 125–126, 129
Gliders, 25
Glue, 24, 27–28, 45. *See also* Cement Glues
Goodyear Pliobond cement, 112
Gouge(s), 136
Grand Prix cars, 54
Grandt Line parts, 94, 96, 106
Graupner winch, 65
Green Stuff, 134
Grinder, 11, 58
Grinding bit, 147
Grinding stone, 146, 159
"Group 12" racer, 173
Half-track, Japanese "Ho-Ha" Type 1, 83, 85–90
Hammer and block, 56
Harris, J. W., solder, 158
Hasegawa, 80
Hathaway, John, 9, 35, 109–110
Hawk kit, 55
Heat sink, 156, 158
Helicopter, 19, 32, 173
Heller kits, 24, 32, 80
Holgate and Reynolds materials, 116
Hot Stuff, 44, 95, 100, 119
Housework brand, 108
Hovercraft, 19
Hull(s), ship
  cross-section of, 72–73
  fiberglass, 65, 135, 159
  finishing, 74–78
  pattern for, 135
  plastic, 63–64
  riveting, 144
  sealing, 67
  solid, 69, 74
  wooden, shaping, 69, 76, 78, 135
Ideal brand, 109
Imai/Scale kit, 7
Injection-molding machine, 159
Jeeps, 80

Jelutong wood, 74
Jig, 126–129, 141, 155
JoHan automobile model, 48
Joints, fitting, 167–168
Kappler wood, 74, 142
Kemtron kits, 93, 99, 100, 106
Knife(ves)
  carving, 136
  hobby, 10, 27–29
  pen, 10
  surgical, 112
  X-acto, 11
Kraft transmitter, 174
K & S brass, 69, 100, 145, 171
Kurton Products, 165–166
LaBelle Woodworking kits, 93
Lacquer, 31, 137–138
Ladder, model, 142
Laminating
  balsa, 140
  plastic, 123
  styrene, 69
  use of, 100
Landscape modelling, 107
Lathe, 144
  chuck, 125, 153
  collet, 125
  plastics with, 123
  turning technique, 15
  using, 11, 135, 151–153
Lesney/Matchbox automobile kits, 48
Lettering, 14, 36
Lighter-than-air craft, 32
Lindberg kits, 24, 32
Lintels, model, 116
Lionel chassis, 2
Liquid aluminum, 61
Locomotive(s)
  boiler, 98, 106, 154, 158
  brass kits, 93
  building, 92
  cab, 96–101, 104, 106
  diesel, 96, 160, 163–164, 172
  frame, 155
  handmade, 106
  metal, 91
  moving, 91
  plastic, 100
  researching model, 101, 106
  steam, 2, 100–101
  tender, steam, 144
  windows, 97–98
Loening aircraft, 3, 9, 34–40
"Lost wax" casting, 92, 99, 104, 106, 159
Lotus cars, 25, 30, 54, 56
Lucite, 123
"Lunar Module" kits, 22, 24
LytLer & LytLer kits, 116
Machines 5, 123, 130, 151–153, 155–156
Machining, 123, 151
Machinist's square, 124–125
Magnuson Models, 165
Mantua/Tyco kits and parts, 96–100, 171
Maserati 5000 GT, building, 35, 54–57, 61
Matchbox cars, 49, 51, 54
Materials, building, 121–176, 179
Mellor consolidation kit, 98–99

Merrimac, 63
Metal. *See* Brass; Copper; Castings
Metal Base Auto Body Solder, 134
Metal kits, assembling, 92
M.G. kits, 49, 50–54
Microflame torch, 157
Micrometer, 4, 11, 124, 151–152, 155
Micro Scale materials, 14–15
Micro-X Peanut Class model, 41
Military accessories, 79, 83, 107
Military vehicles, making, 79–90
Mill(s) power, 11
Milled shapes, 22, 145
Milled sheet plastic, 123
Milling
  machine, 58, 69, 123, 130, 135, 144, 151–152
  technique, 11, 15, 155
Minicraft kits, 32
Minicraft/Hasegawa series, 80
Mini-Tanks, 80
Mirage 111C jet fighter, 9
Model Masterpieces, 165
Model Parts, plastic materials, 36, 95, 124
Model Shipways, 67–68, 78
Model Shipyard, 78
Model T kit, 150
Module, railroad, 91, 166, 169
Mold(s)
  casting, 60–61, 140, 161–163
  one-piece, 163–164
  for plaster, 58–59
  rubber, 56, 106, 161–164, 167
  two-piece, 161–164
Monitor, 63
Monogram kits, 32, 35, 50, 54, 80, 83
Motor(s)
  electric, 67, 172–174, 176
  servo, 65, 67, 174–176
  wind-up, 172
Motorcycle(s), 48, 51, 80
Motorized miniatures, 19, 90
Mountains in Minutes, 166–167
MPC/Fundimensions automobile kits, 48
MRC, 80
MRC/Tamiya models, 32, 80, 82–84
Mullions, 109
Mustang P51D, 33–34, 41
Nail(s), simulating, 140
Nailheads, making, 78
NASA vehicles, 19, 31
National Association of Rocketry (NAR), 31
Nickel-iron, 158
Nickel-silver, 145
Nitrate plastics, 62
Northeastern wood and kits, 22, 63, 93
Northwest Short Line
  chopper, 126, 141–142
  press, 129–130
  punch and die sets, 131
  riveter, 130

INDEX

101 Productions, 6, 109
Our House kit, 108
Paasche airbrush, 12
Pactra paints, 11, 15, 31
Pactra Solar Film, 45
Paint(s)
  acrylic, 31, 138
  aerosol, 11–13
  clear, 15, 31
  enamel, 31, 137
  epoxy, 59, 137–138
  flammable, 11
  flat finish, 12, 14–15
  fuel-proof, 31
  gloss, 11, 14–15, 78
  lacquer, 31
  primer, 134
  spray, 11, 46
  thinner, 143
Painting. *See also* Weathering
  flying aircraft models, 34, 45–47
  railroad models, 34, 92
  rockets, 30–31
  structures, 92, 117, 119, 143
  technique, 11–14, 45
PanaVise vise, 124–125
Panzer MkIII, 82, 84
Parachutes, ejecting, 25, 27, 30
Patterns, master
  making, 33, 135–136, 138, 168
  using, 59–61, 161, 163, 167
"Peanut Scale" flying model, 32–33, 35, 41, 170, 172
Peanut Sport Model, 41–43, 46
Peck-Polymer
  flying aircraft model, 33, 35, 41
  "Gangobie" flying "Peanut Scale" airplane, 170
  nose bearing, 44–45
  P-51D Mustang, 41
PFM vernier calipers, 124
Photo-etching, 147–148, 150–151
Pinewood Derby cars, 54, 56
Plans
  for aircraft, 177–178
  for armored fighting vehicles, 178
  for cars, 178
  enlargement of, 7, 9
  for rockets, 177
  scale, 7, 9, 68, 116
  for ships, 178
  for structures, 142, 178
  for trains, 178
Plaster casting, 58–60, 159, 165–166
Plastic(s)
  armor plate, making, 85–90
  butyrate, 35, 56
  castings with, 167
  cementing, 127, 129
  cutting, 11, 123–125
  expanded foam, 139
  heat-formed, 123, 132, 136
  injection-molded, 54, 62, 100, 132–133, 160
  precision work with, 123–125

Styrofoam, 67
  trains, 91
  vacuum-formed, 58–61, 132–133
  working with, 15, 83, 127–129
Plastic wood putty, 134
Plastruct, 22–24, 36, 69, 124
Platen, 59, 61
Plexiglas, 36, 68, 74, 85
Pliers, 10–11, 144–145, 156
Plywood
  bases, 91
  bulkheads, ship, 77
  for flying aircraft, 140
  scrap, 125
Polks models, 80, 110, 116
Pontoons, 36
Porsche, 163
Precision Investment Associates, 106
Precision Manufacturing Company, 130
Precision Miniatures, 36, 50–51
Precision Scale Company, 93, 106
Primer, 81–82, 137, 139
Probar winch, 65
Pro-Cision "Sun Cruiser," 172
Punching, 11, 129–132
Quality Craft, 93, 96, 151
Quick Silver Putty, 134
Racing car(s)
  body, 57, 138
  electric slot, 174
  radio-controlled, 48, 169–170, 173–176
Radio control (R/C)
  automobiles, 175–176
  flying aircraft, 32, 169
  installation, 175–176
  race cars, 107
  receivers, 41, 174–176
  sailplane, 41
  ships, 65
  tanks, 90
  transmitters, 41, 174–176
Railroad
  cars, 3, 91–93, 163
  control systems for, 174
  equipment, 6
  kits, 92, 108
  materials for, 92
  models, 2–3, 19, 34, 48, 92, 142
  rolling stock, 91–92
  scenery for, 107
  structure, 91–92
  tracks, 91–92, 145
Resin(s)
  casting, 135, 161–165, 167–168
  epoxy, 58, 61, 167
  hardened, 138
Revell models, 24, 32, 35, 170
Rio metal automobile, 49
Rivet(s)
  embossing, 83, 100, 105–106
  heads, 60, 82, 150–151
  impressing, 15
  making, 38–39, 85, 144
  simulating, 2, 126, 130–132

Robot, radio-control, 19
Rocket(s), static
  building, 19–31, 67
  engines, 25
  fantasy, 19–20, 25, 31
  kits, 19–20, 22
  materials for, 91
  painting, 30–31
  parts, 20
  plans for, 177
  researching, 31
Rocket(s), flying
  bodies, 25–30
  booster, 27–28, 30
  "Double Trouble," 25–28, 31
  engines, 25–27
  fins, 25, 28
  launch stand for, 25
  making, 19–20, 24–31
  nose cone, 25, 27–28, 30
  parachute, 25, 27, 30
  two-stage, building, 27–30
Rockwell belt grinder, 147
Romanoff Rubber Company, 160
Rosin flux, 156
Roundhouse, railroad, 107
Rubber
  latex, 166–167
  mold, 167
  RTV, 56, 106, 161–164
  tires, 161
Rubbing compound, 138
Rub 'n Buff paste, 82
Sailplane, 41, 63, 169, 174
Santa Fe Railway, 98–99, 101–104, 106
Saw(s)
  electric, 11
  hand, 11, 74
  jeweler's, 145–146, 167
  jig, 11, 136–137, 140–141, 146
  razor, 93, 141, 145
  saber, 74, 146
Scale
  defined, 4–5
  models, 3–5, 8
  proportions, 4, 8
  structures, 112–113, 117
ScaleCraft, 80
Scenery, 91, 166, 169
SC 488 110-foot submarine chaser, 71–73
Schooner, "Bluenose," 65
Scientific Models, 64–65, 69
Scissors, 10–11, 111, 145
Score-and-break technique, 82–83, 93–95, 125–126
Scratchbuilding, 1–2, 11
  aircraft, 34–35, 40, 44, 139
  armored fighting vehicles, 83
  automobiles, 54
  boats, 135
  in brass, 106
  hovercraft, 19
  locomotives, 100–101, 106
  railroads, 92–93
  ships, 67–69
  structures, 113
  warships, 65
Screwdriver(s), 10, 144

Scribing, 129–130
Sealing rubber, 160
Seaplane, 65
Sears belt grinder, 147
Sequoia Scale models, 142
Shark Racing Bodies, 55
Sheet metal
  cutting, 96, 131, 146–147
  etching, 147–151
  forming, 135
  rivets in, 126, 130
  shaping, 100
Shellac, 78, 135
Sherline machines, 123, 151
Shingles, 107, 117
Ship(s), building
  deck, 67–69, 75–76
  fittings, 63, 69
  frame, 69, 74
  guns, 69, 76
  hulls. *See* Hulls, ship
  keels, 74–78
  mast, 68, 78
  models, 19, 63–78
  parts, 63–69
  planking, 76
  plans for, 178
  powered, 65–67
  R/C, 65, 67
  ribs, 69, 74
  rigging, 63–64, 67–68
  rudder, 78
  sailing, 64–65, 107
  spars, 68, 78
  square-rigged, 64, 68
  stern, 69, 76
Ship(s), kit
  assembling, 67–68
  conversions, 68
  cutout, 109
  plastic, 63–64, 67, 76, 78
  Sig materials, 45, 62
Silastic RTV-31, 161
Silicone, 161, 166
Silver Streak/Walthers freight car kit, 93
"Slot" cars
  body, 55
  chassis, 54, 62
  electric, 169, 174
  kits, 48
  skin of, 58
  track, 170
Soldering
  brass, 93
  defined, 155–156
  flux, 144
  gun, 144, 155–158
  iron, 155–158
  multipart, 157–158
  sweat, 157–158
  tools, 156–157
  torch, 144, 155–157
Solido metal automobile, 49
Spacecraft
  building, 19–31
  domes for, 132
  researching, 31
"Space 1999" kit, 22, 24
Space shuttle, star probe, 24
Spitfire, 33, 35
Spray booth, 13
Sprayment adhesive, 111

INDEX

Sta-Brite solder, 144
Staining wood, 142–143
Standard Baldwin Locomotive Works, 99
Starship, Enzmann, 21, 23–24
Static models
  aircraft, 32–33, 91, 169
  armor vehicles, 91
  automobiles, 48, 54, 91
  fantasy vehicles, 20
  railroads, 91
  ships, 91
  spacecraft, 19–20
Station(s)
  cutout, building, 108–113
  passenger and freight, 114–115
  railroad, 35, 107, 113
Stay-Brite, 156, 158
Stay Clean flux, 156
Steel
  riveted, 6
  rule die, 130–132
  simulating, 145
  soldering, 145
  stainless, 145
Steering system, 173–175
Stencils, self-sticking, 36
Sterling models, 65, 69
"Stick" models, 34, 40–45
Stone, simulating, 116–117, 123
Structure(s)
  model, 107–119
  plans for, 178
  staining, 142–143
  windows for, 92
  wooden, duplicating, 140–141
Structure Company, 142
Structure kits
  in cast plaster, 166
  painting, 92
  wood, 112–113, 116–117
Sturminfanterie Greshetz 33 (StuIG 33), 82–84

Styrene
  clear, 61
  milled, 124
  scribing, 129
  shaping, 123
  sheets, 36, 66, 82, 83, 124, 127, 129
  strips, 36, 69
  surface of, 126
  using, 123, 132
  in vacuum-forming, 62
  windows, 15
Styrofoam plastic, 67
Submarine, 65
Submarine chaser, 70–74
Suncoast kits, 108
Sun Cruiser, 172
Superdetailing
  armored military vehicles, 80, 83
  car bodies, 48
  ship's rigging, 67–68
Super Jet cyanoacrylate cement (Goldberg), 24, 27, 49–50
Suydam kits, 108
Tamiya models, 80, 82
Tank, 79–80, 106–107
Techniques, building, 121–176
Thayer and Chandler airbrush, 12
Thermopylae, 63
3M brand materials, 111, 134
Tin-lead alloy, 155, 159–161
Tire(s), 6, 50
Tissue-covering techniques, 45–47
Tissue paper
  for aircraft, 40–41, 140
  Japanese, 41, 45
  using, 45–47
TiteBond, 44–45, 75, 117
Tix flux and solder, 156–158
Tools
  alignment, 124, 144
  hand, for metal working, 144–146
  high-speed motor, 11
  measuring, 11
  model-building, 10–11
  for modelers, 179–180
  power, 11
  precision measuring, 124, 144
Top Flite Models, Super Monocote, 45
Tracing, 83, 85, 87
Tractor miniatures, 48
Trailer miniatures, 48
Train(s). See Railroad(s)
Transformer, soldering, 157
Trees and shrubbery, kit-built, 92
Trestle(s), 142–143
Trim adjustments, 47, 65
Truck(s)
  body, 62, 129
  kits, 48, 50
  models, 49–50
  wheels, 83, 85
Tyco locomotive, 96, 99
Tyco/Mantua trains, 96, 98, 101
U-control, 41, 170, 173
Unimat 3, 123, 130, 151
Union Pacific diesel locomotive, 160, 163
Vacuum-forming
  at home, 61–62
  machine, 36, 59–62, 132
Varnish, 78
Vehicles, military, 79–90
Vollmer kits, 116
V-2 rockets, 19
Walthers materials, 14, 78, 145, 153
Warehouse, model, 116
Wargames, 31, 68, 79
Warships, 64–65, 67, 136

Waterline models
  dioramas, 107
  scale of, 63, 66, 68, 70
  scratchbuilding, 68–69, 75
  shape of, 64
Water tank wrappers, model, 142
Wax modeling, 138, 159
Weathering
  armored fighting vehicle, 80–82
  flying model rockets, 31
  plastic kits, 23, 34
  railroad models, 92
  structures, 119
Weight, in model engineering, 169–170
Welding, 155, 158
Weld-On cement, 117
Weller soldering gun, 156
Wills Finecast models, 2, 49–50, 54
Winch(es), 65, 67
Windshield wipers, 169
Wold airbrush, 12
Wood
  auto models, 10, 54–55, 57
  balsa. See Balsa
  bass. See Basswood
  castings with, 167
  kits, 92, 107
  locomotive, 99
  for model work, 135–143
  railroad cars, 91–92
  railroad structures, 91–92
  shaping, 11
  sheet, 11
  ships, 63–64, 67–68
  simulating, 117, 123, 126
  staining, 142–143
  strip-, 116, 165
  working with, 15, 112–116
Woodhill putty, 134
X-Acto tools, 11, 12, 95, 145
Zeppelin(s), 32

```
j688.1                    820065
Schleicher
    The modeler's manual.
```

```
j688.1                    820065
Schleicher
    The modeler's manual.




        BELLAIRE PUBLIC LIBRARY
          BELLAIRE, OHIO




                                GAYLORD R
```